Geographic Information Systems in Action

Michael N. DeMers
New Mexico State University

VP AND EDITORIAL DIRECTOR	Petra Recter
EXECUTIVE EDITOR	Jessica Fiorillo
EDITORIAL MANAGER	Gladys Soto
CONTENT MANAGEMENT DIRECTOR	Lisa Wojcik
CONTENT MANAGER	Nichole Urban
SENIOR CONTENT SPECIALIST	Nicole Repasky
PRODUCTION EDITOR	Loganathan Kandan
PHOTO RESEARCHER	Alicia South-Hurt
COVER PHOTO CREDIT	© Pablo Scapinachis/Shutterstock

This book was set in Times LT Std 10/12 by SPi Global and printed and bound by Strategic Content Imaging.

This book is printed on acid free paper. ∞

Founded in 1807, John Wiley & Sons, Inc. has been a valued source of knowledge and understanding for more than 200 years, helping people around the world meet their needs and fulfill their aspirations. Our company is built on a foundation of principles that include responsibility to the communities we serve and where we live and work. In 2008, we launched a Corporate Citizenship Initiative, a global effort to address the environmental, social, economic, and ethical challenges we face in our business. Among the issues we are addressing are carbon impact, paper specifications and procurement, ethical conduct within our business and among our vendors, and community and charitable support. For more information, please visit our website: www.wiley.com/go/citizenship.

Evaluation copies are provided to qualified academics and professionals for review purposes only, for use in their courses during the next academic year. These copies are licensed and may not be sold or transferred to a third party. Upon completion of the review period, please return the evaluation copy to Wiley. Return instructions and a free of charge return shipping label are available at: www.wiley.com/go/returnlabel. If you have chosen to adopt this textbook for use in your course, please accept this book as your complimentary desk copy. Outside of the United States, please contact your local sales representative.

ISBN: 978-1-119-23886-7 (PBK)
ISBN: 978-1-119-22283-5 (EVALC)

Library of Congress Cataloging in Publication Data:

Names: DeMers, Michael N., author.
Title: GIS in action / Michael N. DeMers.
Description: Hoboken, NJ : John Wiley & Sons, 2017. | Includes bibliographical references.
Identifiers: LCCN 2016054687 | ISBN 9781119238867 (pbk.) | ISBN 9781119227373 (epub)
Subjects: LCSH: Geographic information systems.
Classification: LCC G70.212 .D468 2017 | DDC 910.285--dc23 LC record available at https://lccn.loc.gov/2016054687

The inside back cover will contain printing identification and country of origin if omitted from this page. In addition, if the ISBN on the back cover differs from the ISBN on this page, the one on the back cover is correct.

DEDICATION

There comes a time in one's career when they realize that thanking all the individuals who have supported them over the years one by one would likely consume another full volume. This is such a time. To save you reading such an exhaustive list, I therefore dedicate this book to all those who taught me through coursework, through publications and presentations, through conversations, or through providing feedback from my classes. I am particularly grateful to the entire education team at Esri and to the T3G instructors and participants who continue to inspire, to share, and to teach. Finally to my friends, my colleagues, and especially my wife of over 30 years, I thank you for your support and dedicate this book to you.

Preface

Geographic information systems (GIS) have been growing immensely, especially in the past 20 years. A major response to that growth is the development of an ever-growing list of textbooks and learning materials. The textbooks, including my own *Fundamentals of Geographic Information Systems,* have largely targeted an intellectual university undergraduate and graduate student audience. This audience may or may not actually apply the tools of the GIS, using it instead perhaps to manage GIS operations or for later PhD work and eventually teaching it. This unfortunately leaves behind the largest growth market in GIS, that of the community college student, who is much more likely to immediately apply the skills learned in a GIS course. This text is aimed at this growth market.

Community college students, the primary audience for this text, are pragmatic, practical, and impatient with nonrelevant course material. They go to community colleges expecting to gain not just factual and conceptual knowledge but also skills that can be applied immediately. If, however, students learn the skills but lack a basic understanding of the concepts behind the GIS toolkit (as occurs when they use many of the cookbook-style laboratory assignments currently available, even those from the major GIS companies), they will perform tasks only mechanically with little concern for or understanding of the reliability or validity of the result.

GIS in Action aims to overcome these shortcomings for community college students by offering content that not only teaches GIS techniques, the ideas behind them, and how they work, but also—through a series of graded, hands-on content-oriented activities—challenges students to think through what they are doing and why before going on to practical ArcGIS exercises. Hopefully, this deeper understanding and the superior problem-solving skills that students gain from using the text will also make them more valuable employees.

Each of the chapters begins with a set of learning objectives followed by matching behavioral indicators based on action verbs that specify how the learner will demonstrate command of the material. The material is presented in very short readings that attempt to get at the important concepts with minimal text. Text material is then followed with activities that attempt to make the link between the conceptual material and, where applicable, the ArcGIS application of those concepts.

The ArcGIS lessons make reference to creating screen captures to be turned in to the instructor; rather than making the hand-in standard, the idea is to allow the instructor to provide his or her method of evaluating learning. These lessons could be anything from completing practical quizzes to requiring artifacts (such as the screen captures) or even full reports describing a process. This way, the instructor can match his or her own teaching style with the level, interests, and needs of the learners.

The software is available from Esri via the URL http://www.esri.com/software/arcgis/arcgis-for-desktop/free-trial. Some instructors are at colleges that have a site license for their institution and can get access to the software from the following site: http://www.esri.com/industries/apps/education/offers/promo/index.cfm. Complete directions for installing both the software and the tutorial datasets are included on the site. Attention should be paid to this because the instructor might find it easier if every student places her or his data in the same locations on her or his disks. Other faculty have laboratories with the software installed on the computers. If that is the case, it is important that a system administrator restore the hard drives to conditions before data were saved. Otherwise, the software could be set up at different places in the laboratory exercises for each department.

It is important to understand that this text is meant to act as a surficial introduction to the general nature and applications of GIS, not as a comprehensive in-depth course. In the years I have taught GIS, I have found that an awareness of the capabilities of GIS is a better beginning than an attempt to master the software. While software skills are important, they, like the software itself, are ephemeral. At the same time, if the learner can see the larger overview of the discipline and can divide it into relevant topics, the organizational structure itself assists them in knowing where to start when asked to perform GIS operations for an employer. Once that has been discovered, the learner will be able to learn the necessary nuances of the software to accomplish what is required of them.

Organization of the Book

GIS in Action has 10 chapters of varying length, depending on the nature of the material, and are not necessarily linked to a 10- or 15-week quarter or semester. Chapter 2, for example, contains much of the background geography and cartography that the learner needs to know to understand what the GIS field is all about. It is up to the instructor to determine how much emphasis to place on any given chapter and to ascertain how long each chapter could be expected to take. In this way, instructors can focus on learner needs, the time frame available for the class, and any additional material they wish to bring in. The chapters flow as follows:

Chapter 1, "Introduction to GIS," tells the student what GIS is and how it relates to their potential employment. This chapter is important because it provides a context for the learner and answers the big question "Why am I learning this?"

Chapter 2, "GIS Representation" gives the student the fundamental building blocks of GIS, GIS data, cartography, projections, datums, and grid systems. This is a long chapter but provides much of the necessary geographic and computer background to move forward. I would suggest spending some time on this rather than just diving into Chapter 3.

Chapter 3, "Creating and Editing GIS Data," provides the student with conceptual and real experiences getting data into the GIS (ultimately, the first step in GIS) as well as the editing process (a close second step). It has been the observation of many of my university graduates that data input, data editing, and data management are still among the most important concepts in the business.

Chapter 4, "Basic Map Queries," is the next logical step once the student has learned how to input data to be able to formulate queries to retrieve data based on a particular set of questions. This chapter emphasizes how to search for the appropriate data using the built-in GIS query tools.

Chapter 5, "Common ArcGIS Tools," provides an introduction to some of the basic tools likely to be used in ArcGIS analysis. In particular, the primary topics of analysis are overlay analysis, proximity analysis and buffers, summary statistics, and table management.

Chapter 6, "Raster Operations," which has the primary focus on the common tools in Chapter 5, is based on vector data types. This chapter makes a leap to raster both because it needs to be studied in such a focused manner and because it actually contains much of the analytical power for ArcGIS and is often ignored in other texts.

Chapter 7, "Network Analysis," presents another topic often ignored in introductory GIS courses, Network Analysis. Granted that much GIS doesn't involve either geometric networks or network datasets, the beginning GIS student needs to be aware of these tools when they are needed. An advanced topic of location-allocation is also introduced in this chapter both because the learner needs to be aware of it and because they need to be aware that such modeling is performed with the use of network datasets.

Chapter 8, "Surface Analysis," emphasizes topographic surfaces, primarily as they are by far the most used 3-D surfaces in GIS. At this point, the learner is aware of both raster and vector analyses, but there has been no mention of how they can be applied to surfaces. The approach first looks at the Geostatistical Analyst extension in vector and then moves to the raster surface toolset. The reason for not separating the raster surface toolset and moving it to Chapter 6 is that it allows the learner to distinguish between the two approaches to modeling surfaces side-by-side.

Chapter 9, "Constructing Models," focuses on the next logical step in skill level development, training the student to construct complex models from the basic tools they have learned. Modeling really is the culmination of all the previous chapters. Many texts ignore this topic and instead assume the student will somehow make the leap from individual analyses to modeling. ArcGIS ModelBuilder is the focal point for this chapter because the toolset is a good place to begin to learn modeling. It is important, of course, to explain to the learner that modeling can become very complex and is not always something that can be constructed using ModelBuilder.

Chapter 10, "Reporting Results," emphasizes process. Once any analysis has been completed, the success or failure of the task hinges on creating "appropriate" and "quality" output. The important thing the learner should get from this is that performing a powerful analysis is not the final step. Once finished, the client and/or audience needs to have the result of that work produced in a clear, concise, understandable manner. This is often not examined in introductory GIS courses and is included here for that reason.

Key Features of the Book

GIS in Action is based on a practical, hands-on, problem-based learning approach. Each chapter provides opportunities for learners to demonstrate their basic understanding of the material through text-based activities and then to immediately apply what they have learned using professional GIS software (ArcGIS 10). The GIS activities challenge students to engage in their learning through activities that teach them not only how the software works but also how the concepts they have learned through the text are implemented and solved using the software. Moreover, at the end of each GIS activity, students are again challenged to think deeply about what they just did, what it means, how it was performed, and even how mistakes can lead to either software difficulties or incorrect answers.

ACTIVITIES and LESSONS. Two types of activities and lessons are interwoven with the material to support the text-based content. The first, called **Activities**, immediately follows small portions of the text to immediately reinforce the material just covered. Second are the **ArcGIS Lessons**, the text of which is provided both within the document and on the Wiley website. The ArcGIS Lessons allow students to apply what has been learned and to have the hands-on experience needed for future GIS practitioners. Using the ArcGIS 10 software and the data provided with those disks, ArcGIS Lessons make a direct linkage between the conceptual content and the applications so critical to effective GIS practice. The following sections provide a bit more detail about these different types of Activities and ArcGIS Lessons as well as other material found in the text.

Activities

These Activities include a comprehensive set of questions designed to get the students to recall what they have read, to reflect on that content, and to analyze, explain, synthesize, and in some cases, evaluate, the content depending on the nature of the material. These Activities may require additional reading, answering questions, reviewing material already covered, or exercising what is learned to demonstrate basic understanding of the content in short easy tasks. As the chapters progress, the focus moves more from basic understanding to the higher levels of Bloom's Taxonomy. Beyond simply demonstrating the level of comprehension, these activities prepare the student for the ArcGIS lessons that employ this content.

ArcGIS Lessons

The section set of lessons is based on hands-on practical applications of the ArcGIS 10.x software to both reinforce the text-based learning and to demonstrate the practical nature of the concepts as applied inside the GIS software. The exercises are platform dependent (Windows), but ArcGIS will run on a Mac with an Intel chipset using VMWare running Windows or running Windows via Bootcamp. The ArcGIS lessons appear within chapters and, depending on their nature, will focus either on learning the basic GIS skillset or on problem solving to allow students to practice their spatial analysis abilities.

Further Readings and Resources and Key Terms

Each chapter ends with a set of additional readings of books and articles, and some chapters identify materials available for free online. These readings are meant to supplement the material here, not to be comprehensive because that is not the purpose of such a practical, skills-based course.

Contents

3   Creating and Editing GIS Data   43

4   Basic Map Queries   109

5   Common ArcGIS Tools   143

6   Raster Operations   158

7   Network Analysis   180

Introduction to Geographic Information Systems (GIS)

CHAPTER 1

LEARNING OBJECTIVES

Here is the content you will learn in this chapter:

1. The different definitions of GIS.

2. The advantages and disadvantages as well as consequences of different GIS definitions.

3. The nature of geographic questions that GIS is designed to solve.

4. The many disciplines to which GIS tools are applied every day.

5. The types of individuals, organizations, and government bodies that can benefit from GIS.

6. Examples of GIS applications for solving different types of geographic problems.

7. The way GIS knowledge and skills can empower these institutions, organizations, and government bodies.

8. The skills and knowledge needed to fulfill the needs of a geospatial workforce.

9. How geospatial knowledge and skills are incorporated into existing work environments.

10. How the incorporation of geospatial skills and understanding can transform your workplace.

BEHAVIORAL INDICATORS

When you are finished with this chapter, you will be able to:

1. Provide up to five unique definitions of GIS.

2. Explain the similarities and differences among up to five unique definitions of GIS. Define what limitations a "software only" definition places on how the discipline of GIS functions and benefits the organizations that incorporate it.

3. Provide a list of nine specific types of geographic questions GIS is designed to answer.

4. List and describe at least five different disciplinary specialties in which GIS is employed and the general tasks that GIS addresses within them.

5. Provide concrete examples of specific organizations (e.g., companies, government bodies, nonprofit organizations) that employ GIS and describe one or more applications that they perform on a regular basis.

6. Provide a list of at least five concrete examples of geographic applications for solving different types of problems.

7. Based on concrete examples of how organizations use GIS, describe how GIS empowers each organization to do what it normally does better, faster, more efficiently, or with less cost. Alternatively, describe tasks that GIS allows organizations to perform now, that they could not in the past and explain why this is so.

8. Describe the different kinds of GIS skills and knowledge that are in demand in the workplace.

9. Provide at least five concrete examples of how GIS knowledge and skills can impact your ability to obtain employment in organizations that are not normally considered "geospatial" in nature.

10. Explain how the incorporation of geospatial skills and competencies might improve the speed, quality, efficiency, or profitability of an organization that is not currently considered a "geospatial" company.

Chapter Overview

This chapter will answer the big question: Why am I learning about geographic information systems (GIS)? They have become integrated into so many facets of our lives that their influence has often been taken for granted. But like many things in our world, the commonplace tools and skills are often the most critical. This chapter provides a brief working knowledge of what GIS is, what problems it can solve, how it can empower organizations and you personally, and what skills you will need to benefit from GIS. You will examine real-world applications provided by business, industry, government, and nonprofit organizations. These examples will show you how GIS can improve the speed, efficiency, quality, and in some cases, profitability of the products and services these organizations produce. While you go through this chapter, think about the kinds of GIS-related employment you are currently involved in or wish to have. Think about how you or your organization can benefit from the use of the empowering technology, the knowledge, and ideas that drive GIS.

What Is GIS?

As a student of GIS, it might seem odd that there is a section of this chapter devoted just to defining GIS. GIS is far more complex as an idea than it is as a technology or even as a software package. The underlying disciplines related to GIS are now collectively called geographic information science and technology (GIS&T). This term implies that a major component of the definition of GIS has to do with the science that drives the GIS. The term also includes technology, and this says that there is a technological aspect to GIS. In fact, there are many different aspects and many different views of GIS depending on how the software is used; the context in which it is considered; and whether it is thought of as a tool, a conceptual framework, or an entire discipline. The following sections will examine these aspects individually.

Database Definition

GIS users who think primarily about the ability of GIS to hold lots of data for storage and eventual retrieval might be observed calling the GIS database itself the GIS. From their perspective, this definition is complete because they focus on how the database is created; how it is stored and organized; what it contains; and how it might be kept safe, remain accurate, and be timely and useful. In many cases, these users are not really defining GIS so much as focusing on that aspect of the GIS with which they are both familiar and concerned.

Within this useful but limited definition, there are many aspects of GIS that are critical to the overall functioning of GIS. Among the most prominent of these is that the user must have the right data, they must be for the correct area, they must be safe, the user must have ready access to them and must have data descriptions that are clear to the user, and the data must be accurate. Database administrators are often the most concerned about GIS as database and therefore pay particular attention to what happens when a database is created, when and how it is accessed, and particularly when it is changed—especially when more than one person has access to the same database.

So from the GIS database administrator perspective, this is an operational definition, but there is one major drawback to the use of this limited definition. The limitation is that many GIS practitioners do more than manage the database. As a result, when a practitioner (e.g., a GIS applications person) and a database manager are conversing about GIS, they may very well misinterpret each other's meanings. A classic example is when someone says he or she has this fabulous GIS and you respond by asking what it does. You will frequently be met by a look of consternation because the database administrator focuses on the existence and maintenance of a clean, orderly, complete database and seldom anything more. The next definition includes all of these factors and details some of the missing parts that the database definition does not include and for which software users might be most concerned.

Software Definition

Perhaps the most commonly used definition of GIS today is that it is a piece of software designed for the input, storage, editing, retrieval, manipulation, analysis, and output of geographic data (Figure 1-1). While this seems simple as a definition, it actually has several aspects that need to be examined. Look at these one at a time.

- **Input.** Much GIS data are not in a form that the computer can use. These may include analog maps (hard copy maps usually on paper or mylar), gazetteers (geographic data stored

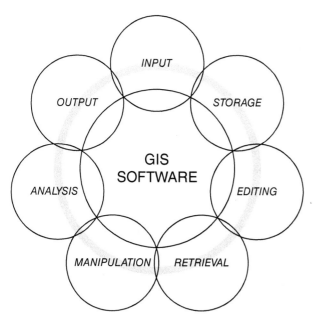

FIGURE 1-1   Components of a fully functional GIS software package.

as tables and lists), environmental field data, economic data, census data, survey measurements, and many other types of data that are not in digital (computer-compatible) form. The software has to be capable of incorporating these data not only in a form that the computer can store and manipulate but also in map-compatible form including all the appropriate locational information, projections, grid systems, and associated descriptive and symbol information. You should also note that the term *data* is purposely used here instead of information because information is most accurately defined as the outcome of some form of data manipulation. In short, information is data in the context of some question or query you pose of the software.

- **Storage.** Once the data are in the GIS, it is important that the software is able to save them so that they are there when you need them for querying or analysis. As simple as this sounds, it also means that the software has to be able to recognize available formats in which the data are stored and within which they can be edited, updated, and corrected.

- **Editing.** The ability of the software to edit data is critical. Correct data are essential to proper analysis. They must be in the correct position, have the correct categories and values, and be timely enough for a given project. GIS software contains a variety of automated and manual techniques that allow the user to ensure that the right data are in the right place.

- **Retrieval.** To perform analysis and to edit and update GIS data, the software must also be able to retrieve the portions of the database from the database that the user requires for a given task. More than just being able to load the entire database, the software must be able to target and select specific pieces of data for these operations. The software must also be able to do this efficiently and with no corruption of the database or any of its parts.

- **Manipulation.** No software definition of GIS would be complete if it did not include the ability to manipulate the data contained in the database. Manipulation has to do with grouping, sorting, geocoding (adding actual locations to the data), and adjusting to changes in map configurations (more on this in Chapter 2). Think of manipulation as preparation for more involved analysis.

- **Analysis.** The real power of the GIS is its ability to analyze spatial data. The analyses include such things as counting; measuring distance, shape, volume, and configuration; comparing individual pieces of data or whole maps; calculating missing values; performing statistical analysis; determining visibility in topographic surfaces; performing basin fill and movement of liquids; and literally thousands more individual operations and combinations of operations. The GIS is able to combine multiple analytical techniques into large, repeatable models that can answer very complex questions.

- **Output.** Without the ability to output information from the analysis of data, the GIS is of little use. I emphasize the word *information* again to remind you that the output is usually the result of answering a question and is now considered to be in the context of that question. Most GIS software is able to produce a wide variety of map types, each with a different purpose in mind. Modern GIS also produces 3-D images, animations, fly-throughs, road logs, and charts, and a host of electronic output including, believe it or not, automated phone calls.

In terms of consequences, you are reasonably safe with this definition because it is a commonly accepted, although somewhat limiting definition. Most GIS practitioners are comfortable with using the term *GIS* to refer to the software but are often aware that a fully functional GIS relies on considerably more than the software. While they may not say it, practitioners often recognize that the "S" in GIS stands for system and that all systems are made of component parts (subsystems), only one of which is the software itself.

System Definition

As you can see, the often-used software-only definition has a lot in it, including all the components of the database definition. It seems to cover many aspects of the software as well. Stop and think for a moment, however, about all these aspects and how they work. If you look at the very first one, input, there are some things you must consider. The following list outlines just a few of them.

- Exactly what are you going to input?

- Who is going to perform the input functions?

- Where are you going to find the data?

- In what format do the data exist?

- Who is going to provide the data?

- What instruments do you need to input the data?

- Will you have to pay for the data?

- Do GIS personnel require training?

- Do you need to reference the data provider when you use its data?

Within that list for just that first factor (input), you start to see some of the things that a software only definition missed. There are people involved in the potential sale of the data, some involved in the data input, and still more involved in the training of the input specialist. These people rely on hardware to input the data, not to mention the storage, retrieval, manipulation, analysis of the data, and output of the results. The people who work in the GIS industry often work for organizations, and these organizations become part of the larger definition of GIS. The organizations might be where you work, or it might be your client, or even a vendor of the data, training, software and/or hardware that supports your operations (see Figure 1-2).

The system definition of geographic information systems, while more inclusive than either the database or the software definition, is less frequently used in common practice. When you are talking about the industry, your profession, or the activities in which you are engaged, you might very well be thinking about GIS from a system perspective. You may also find yourself using the software definition when you are concerned with algorithms, problem solving, and other internal

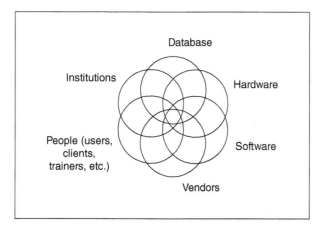

FIGURE 1-2 The system definition of GIS includes the database, hardware, software, vendors, people, and institutions.

operations. The definition you use, then, is context dependent and changes based on the nature of your work and conditions within which you are working. Academics, for example, recognize these multiple definitions and include geographic information science or geographic information science and technology (GIS&T) when including the theoretical concepts and technological advances that drive the advances in the industry.

ACTIVITY 1-1   DEFINING GIS

In this activity, you will have the opportunity to demonstrate your mastery of the basic definitions of GIS. A working knowledge of the different context-sensitive definitions of GIS will allow you to communicate effectively with other GIS professionals. The ability to understand GIS professionals that work in different parts of the industry will allow you to be understood. Of equal importance is that your colleagues will have confidence in your GIS background and training.

Scenario

You are at a GIS conference and are having a conversation with someone whom you discover was a database management person at a business and is now being asked to learn about GIS because the company plans to move toward incorporating GIS into their operations. As you converse, you discover that you seem to be talking about completely different things when you talk about GIS.

1. In the space provided, list and describe five definitions of GIS you have discovered on the Internet so you will be prepared to talk to folks at the conference intelligently.

   _____

   _____

   _____

   _____

   _____

   _____

   _____

   _____

   _____

2. Given this person's background, what do you believe will be her understanding of the definition of GIS?

   _____

   _____

   _____

   _____

3. What are the two primary missing components of that definition?

   _____

   _____

   _____

   _____

4. Describe to this person, in your own words, how the meaning of GIS is broader than the one she understands by providing examples of the importance of the database, the analysis, and the institutional components of a more complete definition of GIS.

   _____

   _____

   _____

   _____

5. You meet a software vendor who tells you that he has the best "GIS" on the market. Rather than disagreeing with him, simply understand his perspective but note below the limitation of the "software only" definition.

   _____

   _____

   _____

   _____

Further Exploration

Go onto the Internet and use your favorite browser to gather a collection of GIS definitions. Evaluate each definition for completeness and consider how each definition might relate to the author's particular bias or professional background. When you have finished, share these with your classmates. Your instructor might even assign this as a discussion activity.

What GIS Does

In the previous section, you learned that one definition of GIS focuses on the software. You also discovered that the software definition is possibly the most prevalent. The reason for this has much to do with the focus of GIS as a tool for analysis that so many GIS practitioners perform regularly. There are many forms of analysis a GIS can accomplish from simple to complex. To understand the real power of GIS, it is best to look at what types of questions it is designed to answer rather than providing an exhaustive list of functions. The next paragraphs group these questions by type.

Questions Regarding Individual Geographic Objects

Every geographic object occurs somewhere in geographic space and its location is often important to the GIS analyst. Typical questions about individual locations of objects include:

- Where is the object located in absolute coordinates? *This piece of information allows many other questions to be asked regarding relative distance.*

- How much space does the object occupy? *Knowing how much space a geographic object occupies can be used to determine value (e.g., the size of an ore deposit) and relationships with other objects (see spatial relationships below).*

- What are the descriptions or values associated with the object? *The descriptive information of individual objects can be used to selectively search for them (e.g., to find all "urban" polygons).*

- When were the object's location, extent, or descriptive information collected (temporal information)? *These characteristics can also include things like when the object first appeared in that location. For example, if a crime occurred at a particular place, when did it occur?*

- Has the object's location, extent, or attributes changed through time? *A good example of this might be asking questions about whether a fire, disease, or insect infestation is increasing in the area or not.*

Some might recognize these as questions one might ask of a traditional map, but as the volume of data increases and as the questions become more complex, the GIS software becomes invaluable for such repetitive tasks.

Questions Concerning Multiple Geographic Entities

More difficult questions arise when we consider relationships among two or more entities. For instance, we can ask:

Questions about Spatial Relationships

- Do the entities contain one another (e.g., does a polygon representing a lake contain an island)?

- Do the entities overlap (e.g., does a study area polygon overlap a grassland polygon)?

- Are the entities connected (e.g., do two roads come together at an intersection)?

- Are the entities situated within a certain distance of one another (e.g., is a road line within 200 feet of a hospital polygon)?

- What is the best route from one entity to the other? (e.g., what is the shortest route from your home [a point entity] and a store [a point entity])?

- Where are entities with similar attributes located (e.g., where are all the urban polygons located)?

Questions about Attribute Relationships

- Do the entities share attributes that match one or more criteria (e.g., find all forest polygons that are larger than ten hectares and greater than one hundred miles from the nearest interstate highway)?

- Are the attributes of one entity influenced by changes in another entity (e.g., as the value of residential land polygons increases, does the average size of home increase)?

Questions about Temporal Relationships

- Have the entities' locations, extents, or attributes changed over time (e.g., has the extent of forest insect infestation increased from 1970 to 1980)?

Geographic data and information technologies are very well suited to answering moderately complex questions like these. GIS is most valuable to large organizations that need to answer such questions often.

ACTIVITY 1-2	WHAT GIS DOES

In the following activity, you will have the opportunity to demonstrate your mastery of the capabilities of the GIS to do analysis. Select some maps and aerial photographs you happen to like from your searches on the web and provide at least two examples of each of the following. You may have to use several maps to complete this activity. Your instructor will likely require you to attach a printout or screenshot of your map objects.

1. Absolute locations (in latitude and longitude) of two point objects (depending on scale, these could be towns, wells, mines, benchmarks, or many others).

   _____

2. Approximate length, width, and area of an area object on a map or aerial photograph (e.g., a football field or baseball diamond, farm section).

   _____

3. From a map of land use, such as vegetation or other categories, list some of the names you encounter for these categories.

   _____

4. From a U.S. Geological Survey (USGS) topographic map, locate the date the map was produced and list it here.

   _____

5. Locate maps or aerial photographs of an area from two different time periods and indicate one major change you can identify (e.g., more urban, damage from a storm or landslide).

   _____

Thought Question: That was pretty easy, right? Now keep in mind that the computer needs to duplicate this process. That's where the power of GIS to keep track of all of this comes in handy.

6. List nine different types of questions that GIS is designed to answer rather than the five basic ones you listed previously.

   _____

   _____

   _____

   _____

   _____

   _____

   _____

   _____

   _____

Further Exploration

Consider your answers above and see if you can come up with two or three concrete examples of each question type. This will get you started thinking like a GIS analyst.

The GIS User Community

As a GIS practitioner, you will soon discover that you are part of a large and growing user community composed of everything from suppliers and programmers to vendors and users. The focus of this chapter is primarily on the largest of these groups—the GIS user community. An awareness of the many aspects and communities of GIS will allow you to know whom to turn to when you need an applications programmer or when you need to design a **geodatabase** (a proprietary database type used by Esri in its ArcGIS software) for a new type of application or when you wish to share data from one application to another. Knowledge of the uses for each discipline and the many different models they employ may also suggest unique and often innovative uses of the many algorithms included in your GIS software. One of the really nice things about the diversity of the user community is that each presents a unique set of challenges and subsequent solutions to often difficult problems. The cross-pollination of ideas from discipline to discipline is one reason that large international user group meetings attract many thousands of users as they share their ideas.

Disciplines Employing GIS

As you can see from Figure 1-3, there are many basic categories of GIS communities and disciplines. The organization into the ten categories in the figure is by no means exhaustive, nor is it the only one that you could find in the literature. Still, it's a reasonable and relatively short list and is easily accessible via the Esri website.

A quick dissection of the groupings is a useful exercise if for no other reason than it allows you to determine into which general category your own applications might fall. The dissemination of GIS is becoming invasive—so invasive that by the time you read Figure 1-3, it is entirely

Industries

Business
- Banking and Financial Services
- Facilities Management
- Insurance
- Media and Press
- Real Estate
- Retail

Defense and Intelligence
- Defense and Force Health Protection
- Enterprise GIS
- Geospatial Intelligence
- Installations and Environment
- Military Operations (C4ISR)

Education
- Libraries and Museums
- Schools (K–12)
- Universities and Community Colleges

Government
- Federal, State, Local, Gov 2.0
- Architecture, Engineering and Construction (AEC)
- Economic Development
- Elections and Redistricting
- Land Administration
- Public Works
- Surveying
- Urban and Regional Planning

Health and Human Services
- Public Health
- Human Services
- Hospital and Health Systems
- Managed Care
- Academic Programs and Research

Mapping and Charting
- Aeronautical
- Cartographic
- Nautical
- Topographic

Natural Resources
- Agriculture
- Climate Change
- Conservation
- Environmental Management
- Forestry
- Marine and Coast
- Mining
- Oceans
- Petroleum
- Water Resources

Public Safety
- Computer-Aided Dispatch
- Emergency/Disaster Management
- Fire, Rescue, and EMS
- Homeland/National Security
- Law Enforcement
- Wildland Fire Management

Transportation
- Aviation
- Highways
- Logistics
- Railways
- Ports and Maritime
- Public Transit

Utilities and Communications
- Electric
- Gas
- Location-Based Services
- Pipeline
- Telecommunications
- Water/Wastewater

FIGURE 1-3   A list of GIS community types supplied by Esri on its website.

likely that many additional applications will be added to the list. In some cases, there may even be whole new communities of users. Many—perhaps most—of you are studying GIS because you intend to be part of one of the current or future communities of GIS users. It is impossible to detail for each of your individual uses how GIS will affect what you do currently. As technology, data availability, and different circumstances drive GIS in your own industry, take the time to keep abreast of how these changes might impact your career. Just as a practical matter, I will now provide a few examples of users and their evolution.

Examples of Users

As you just read, there are many communities and far more individual users of GIS. Each of you is likely to be working with some subset of the industries in Figure 1-3 and you will discover that each industry has multiple communities (subsets) of users who may specialize in one field or another. One quick example would be the water resources industry. Within that group are many different focus groups, and each has its own issues and community (Figure 1-4).

What Questions Users Ask of GIS

Earlier in the chapter, you examined different types of basic questions as well as more complex questions a GIS can answer. Searching the Internet will allow you to find an enormous amount of specific information about how both simple and complex questions are applied by different industries. Let me provide you with a quick example to get you started.

The very first GIS question listed in this chapter was about locating objects. This means the GIS software needs to help you identify where things are on the map. Zoom in to this example from the Internet that illustrates how police know the exact location of crimes because the data

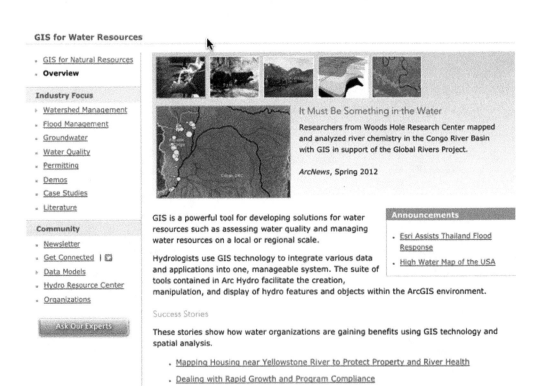

FIGURE 1-4 Screenshot of the Esri website looking at the GIS for the water resources industry.

are clearly indicated both graphically and in specific location data in the GIS data tables. Notice the descriptions of the crimes—the crime types—are also indicated by different symbols. A more sophisticated analysis you will learn about later created those patches of color indicating **hot spots** or **clusters** of crime. Another question this clearly demonstrates is how change has taken place in crime based on the housing foreclosure crisis in Oakland, California. This is a classic example of how the GIS can assist with analysis of temporal change.

How GIS Empowers These Industries

No matter what the industry, there are major general types of advantages to the use of GIS as long as spatial data are used in everyday operations. The shear number of operations (e.g., map analyses) that are capable with a computer and the massive savings in time often translate directly into savings in costs, increased profits, and greater efficiencies in the workplace. In some cases, even entirely new business ventures and new products develop where operations allow analysis of volumes of data previously beyond manual approaches. Businesses also benefit from the perception by their clients of being advanced and familiar with modern technology. That perception alone can increase the client base, increase visibility and use of industry services, and often increase profits. Digital data archiving alone allows access to data that may have been long ignored. The improved quality and increased quantity and variety of digital products developed through GIS is another way GIS empowers organizations.

The addition of GIS as a technology often forces organizations to reconsider inefficient workflows and improve the overall organizational efficiency. While the introduction of GIS can cause some conflicts in organizations, especially among nontechnical personnel, wise GIS implementations through effective design strategies can actually improve morale and promote a sense of belonging and esprit de corps.

ACTIVITY 1-3   THE GIS USER COMMUNITY

In the following activity, you will have the opportunity to demonstrate your mastery of the industries and communities that use GIS. You will also be given an opportunity to relate the types of work these groups do to the kinds of analysis of which GIS is capable and how these techniques can benefit and empower the industries that employ GIS.

Go to the website http://www.esri.com/ and click on the "industries" tab on the top left, right next to the Esri logo. Select two industries, at least one of which is of interest to you, from among the many.

1. Name the two industries you have selected and next to each list the different industry focus groups.

_____

_____

_____

_____

_____

2. Select at least one of the industries you identified in the first activity, in particular one that you are interested in. On the website for that industry, go to community and click on the Get Connected button. Describe the types of information you find there.

_____

_____

_____

_____

_____

3. List and describe at least five of the disciplinary specialties from within the two industries you selected. Next to each industry, indicate the general tasks that GIS addresses for it.

_____

_____

_____

_____

(*continued*)

ACTIVITY 1-3 (CONTINUED)

_____

_____

_____

4. Select from your two industries the names of organizations that belong to the user community (e.g., company names, nonprofits, government agencies).*

_____

_____

_____

_____

5. Go back to the list of questions GIS is capable of answering. Keep them handy. Now do a WebQuest (search the Internet) by picking keywords related to two or more industries and examine examples of GIS work. The keywords might be selected or modified from the list of questions you have at your fingertips. As a hint, for the example from page 10, I used the following terms in a Google search ("locating crime" "GIS") and then selected

images so I could visualize the output from GIS. By clicking on the image, I could also navigate to the website so I could learn about what was happening.

Now find five examples of real GIS applications and list the URL of the image for each. Next to each of these, describe at least one question that was asked in the application (be specific; e.g., this GIS application showed the change in crime locations through time).

_____

_____

_____

_____

_____

_____

_____

_____

These companies may very well be your future employers.

The GIS Skill Set

As a student of GIS, you might find it odd that there is a section of this chapter devoted to the skills you will learn in GIS. Unfortunately, many employers, especially those new to GIS, have little idea what specific skills they need when they advertise for open positions. You may see something as vague as, "Three years of GIS experience" to as specific as "Specializing in the development of enterprise geodatabases." Unfortunately, this doesn't give you much help in preparing for a career in GIS. It is also reasonable to assume that many of you will select different career paths in GIS from applications to modelers, to designers to programs, and many more. Fortunately, an ambitious graduate student at the University of Washington, Michalis Avraam, has created a blog with just the information you need. The URL for his blog is https://tinyurl. com/zkcteqw. Michalis divides these skills into five sets: GIS Skills (including basic spatial understanding), Programming Skills, Database Skills, Project Management and Design, and Other Skills. The relative importance of these five groups of skills varies both based on the kind of organization for which you hope to work and the ever-changing needs of the industry. Take some time to peruse Michalis' blog and read the specific examples of how the skills are exactly translated into the workplace.

Beyond these specific industry-wide and industry sector technical skills, the U.S. Department of Labor has developed a **geospatial competency model** (https://tinyurl.com/z2ly5og) that not only indicates the need for technical skills but also shows the importance of personal, academic, and workplace skills (Figure 1-5).

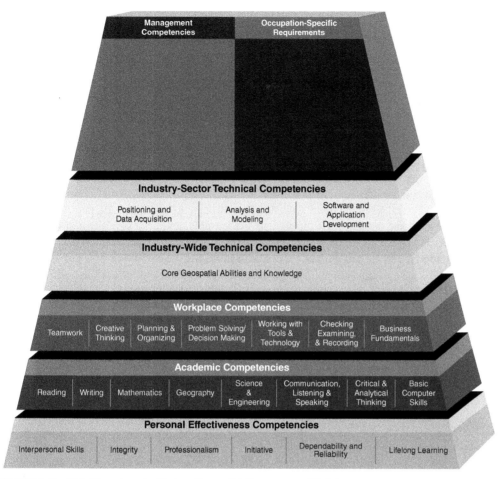

FIGURE 1-5   U.S. Department of Labor Geospatial Competency Model.
https://tinyurl.com/z2ly5og

ACTIVITY 1-4 | THE GIS SKILL SET

In the following activity, you will have the opportunity to demonstrate your mastery of the GIS skill sets that will assist you in employment.

1. What are the five categories of skill sets listed in Michalis' blog?

_____

_____

_____

_____

_____

2. From that blog, your text, and any outside reading, list five concrete examples of how specific GIS knowledge and

skills can help you in obtaining employment in non-GIS organizations.

_____

_____

_____

_____

_____

_____

_____

_____

_____

Incorporating GIS Knowledge and Skills into Your Work and the Workplace

Depending on what career path you choose within or outside the GIS industry, your technical GIS skills will no doubt improve your chances of employment and also improve, if not transform, the workplace. Your spatial understanding, ability to work with different data models and data structures, knowledge of computer science and programming, modeling and data management skills, ability to move between computer platforms, and project management skills are all transferable to nearly all GIS-related industries and a great many non-GIS industries. As you move up in the industry, it is often likely that you will move into the institutional design and strategic planning aspects of the industry. There are instances when companies do not currently use GIS but whose businesses constantly rely on spatial information. Some examples are grocery stores that need to understand their nearby customer needs, hospitals that need to route ambulances and have enough beds for patients in their area, and drug stores that are attempting to make sure they have the kinds of pharmaceuticals for a neighborhood that has a large concentration of elderly. Places that deliver pizza that want to improve delivery times for their pizza will benefit from the routing capabilities of GIS. Colleges and universities can use GIS to recruit the best students. Mapping companies can produce considerably more maps and map updates digitally than they can manually. These are just a few examples. Perhaps you can think of many more.

ACTIVITY 1-5 THE GIS SKILLS AND KNOWLEDGE AT WORK

In the following activity, you will have the opportunity to demonstrate your mastery of the GIS skill sets as they apply to improving the workplace.

1. Select some form of business and describe how GIS tools might be applied to improving profitability. The business type must be real (e.g., liquor store, grocery), but the specific store name is not necessary.

2. Select some form of nonprofit organization and describe how GIS tools might improve its ability to perform its mission.

3. Explain to your friend how GIS might make government (local, regional, county, federal, etc.) more efficient and more responsive.

ADDITIONAL READING AND RESOURCES

Mitchell, Andy. *The Esri Guide to GIS Analysis*. Vol. 1, *Geographic Patterns and Relationships*. Redlands, CA: Esri Press, 2001.

INDUSTRY PROFILE

United States Department of Labor, Employment and Training Administration. http://www.doleta.gov/brg/indprof/geospatial_profile.cfm

KEY TERMS

cluster: An agglomeration of geographic objects so that they occur in close proximity to one another leaving empty space surrounding them.

geodatabase: A proprietary database model that stores, queries, and manipulates spatial data. Beyond a mere geographic database it stores geometry, spatial reference, attributes, and behavioral rules for the data.

geospatial competency model: A collaborative effort of the Employment and Training Administration (ETA), the GeoTech Center, and industry standards that provides guidance for the skills and knowledge needed by today's geospatial technology professionals.

hot spot: An area on a map that displays concentrations, or relatively high densities, of events or geographic features.

CHAPTER 2

GIS Representation

LEARNING OBJECTIVES

Here is the content you will learn in this chapter:

1. The importance of geographic data in GIS activities.

2. The spatial dimension critical to understanding GIS problems.

3. The temporal dimensions of GIS required for modeling the changes of spatial patterns through time.

4. The five levels of geographic data measurement used in GIS datasets, their structures, and models.

5. The five geographic data measurement levels with practical examples.

6. The dimensionality of the geographic objects used in GIS.

7. Geographic scale and its effect on geographic dimensionality and data analysis.

8. The symbology of all types and dimensions of geographic objects, how they are treated inside the GIS, and how they are applied to final mapping.

9. Basic referencing systems (coordinates and datums) commonly used in GIS, their relationships to the location of geographic objects, and the ability of the GIS analyst to measure geographic features using them.

10. The two primary computer data models—vector and raster—used to convert geographic features and data into something the GIS software can recognize and the analyst can manipulate.

BEHAVIORAL INDICATORS

When you are finished with this chapter, you will be able to:

1. Discuss the unique nature and importance of GIS data to GIS operations.

2. Discuss the difference between the spatial and the temporal dimensions of GIS data and give examples.

3. Using GIS software, demonstrate your ability to recognize the five levels of data measurement.

4. List, describe the four types of geographic features or objects and, using GIS software, provide two examples.

5. Identify and/or draw the representation points, lines, areas, and surfaces in raster and vector data models.

6. Using real GIS data examples, explain the functional differences between raster and data models (e.g., storage, modeling capabilities, accuracy, display).

7. Explain the change of scale and how it affects dimensionality and data analysis.

8. Show the relationship between symbols and geographic objects with special emphasis on dimensionality.

9. Using GIS software, find specific locations using the basic coordinate systems and make simple measurements.

10. Identify different dimensions of geographic features in each of the two basic GIS data models (raster and vector).

Chapter Overview

Chapter 1 introduced you to the general idea of what GIS is and its strong linkage to the concepts and ideas of geography, especially place, distribution, and areal correspondence. You will now learn about how real geography is selectively transformed to very basic elements, turned into computer-compatible form, and ultimately converted to the familiar map format after the computer software analyzes it. As you go through this chapter, you will learn about the dimensions (represented as D as in 3-D) in which geographic features exist and how they can be named or have different levels of numeric values assigned to them so that they can be coded for use inside your GIS software. The chapter will identify a couple of common referencing systems used in GIS and give you an opportunity to work with them for locating and measuring geographic features. You will see and experience the two primary ways that geographic features can be simplified and abstracted so that they can be put inside the GIS software. Once you are familiar with those two abstractions—called *data models*—you will be given an opportunity to evaluate the advantages and disadvantages of each.

Nature of Geographic Data

Necessary components of any GIS are the data (facts, numbers, and statistics collected for description or analysis) upon which the software performs its many calculations. The specific kind of data that GIS uses are called *geographic data* or **geodata** (from the Latin word *geos,* meaning earth) and deal with objects and their location on the Earth (or locational information attached to them). Each piece of geographic data has an identifiable position or location that can be pinpointed on or linked to a specific place on the surface of the Earth. The GIS software must have some way of recording these locations so that it can perform calculations on both the data and on their locations. You saw in Chapter 1 that the way geographic data are spread out across geographic space is known as their **spatial**. This link between data and their positions on Earth is important because it is these individual locations and the collections of locations (called **distributions**) that are at the heart of geographic thinking and spatial analysis that drives the GIS.

As you learned in the first chapter, the Earth is filled with geographic features and events that individually occur at specific locations and times. A geographic feature, event or data may be **persistent** (have existed for a long time), such as a city, or it may be very short lived, such as a bolt of lightning striking a tree. You are now aware that individually, these features and events occur in specific places and that collectively they occur in groups called *distributions.* Distributions create observable patterns that vary in size and shape but also vary based on what they are composed of. For example, the distribution of a crowd of people waiting to get into a concert is decidedly different than that of bears in a forest in a number of ways. There are differences in the number of individuals, in density of individuals, and in how the individuals are placed relative to each other. Not only are the distributions quite different, but also the forces that created those differences are also different. The concert crowd is a result of anthropogenic forces including decisions to gather at the specific location at the same time, the shape and size of the gathering place, and even the type of event that brought people there in the first place. The distribution of bears is a natural phenomenon dictated partly by availability of food, partly by availability of shelter, and partly by other factors, such as random chance. This relationship between distributions and their causes is a basic geographic principle used in GIS analysis often referred to as "pattern and process." To understand the causes of patterns, you must first be able to observe and recognize the patterns and then begin to describe and characterize them.

Types of Distributions

There are three basic patterns of distributions: uniform, clustered, and random. The next paragraphs allow you to look at each of these in turn, along with examples, and explore which forces or processes that created them are probably natural or anthropogenic.

Uniform Distribution

Some things on the Earth's surface are distributed such that they are at nearly identical distances from one another, giving them a **regular** pattern. The identical spacing of trees in an orchard (Figure 2-1) or the even spacing of rows in row crops, such as onions, potatoes, and corn (Figure 2-2) are examples of uniform distributions.

    The uniform distribution isn't often a natural type of phenomenon. In most cases, a person was involved in purposely placing these features so they are well ordered. The orchard trees in Figure 2-1 are planted at very specific distances from each other as are the rows of crops in Figure 2-2. What this means is that the pattern (uniform in this case) is related to a process (planting, in this circumstance). A GIS analyst will be able to determine mathematically whether

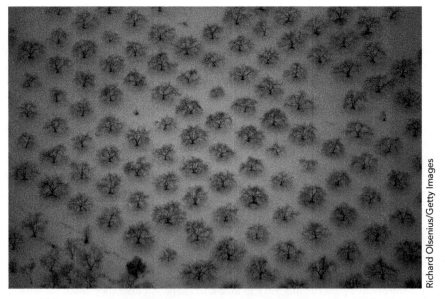

Richard Olsenius/Getty Images

FIGURE 2-1 This apple orchard in Michigan demonstrates uniform distribution. Notice that each tree is the same distance from each of its neighbors.

nrcs.usda.gov

FIGURE 2-2 Typical row crop configuration demonstrating the uniform distribution of each row of the crop in the field.

features like trees are uniformly spaced and, as a result, be able to conclude that humans had something to do with the spacing.

Clustered Distribution

Some features on the Earth's surface tend to be far less uniform than orchards and crops, but they are not completely random, either. Think for a moment about how industries tend to occur in concentrated locations within a city or how people gather at bus stops or shopping malls. They are not uniformly distributed; rather, they occur in quite noticeable groups or **clusters**. Certain types of calls for emergency service also occur in clusters (Figure 2-3). In fact, crime mappers have specific names for these. They call them **hot spots**, and many GIS packages have programs that can find these hot spots and display them on maps.

Clustered distributions can be caused by a variety of things, depending both on what is being grouped and on what forces are acting on those things. Leaves cluster in the wind as fences and shrubs trap them; people cluster near restaurants when they are hungry; wild animals cluster around water holes when they are thirsty; and thieves cluster where there are likely victims or things to steal. Again, the GIS analyst is able to define these clusters and display them on the map. Once displayed, these patterns can easily be compared to other variables that display similar patterns to assist the analyst with determining possible causes for the patterns.

Random Distributions

Somewhere between the even spacing of uniform distributions and the clumped nature of clustered distributions is one called *random*. Random distributions are more chaotic and have a less defined pattern. They may have no identifiable pattern at all although they always seem to trend toward either uniform or clustered. Random distributions might be seen in the disorganized locations of dandelions in a backyard or in the uneven distribution of trees in a forest (Figure 2-4).

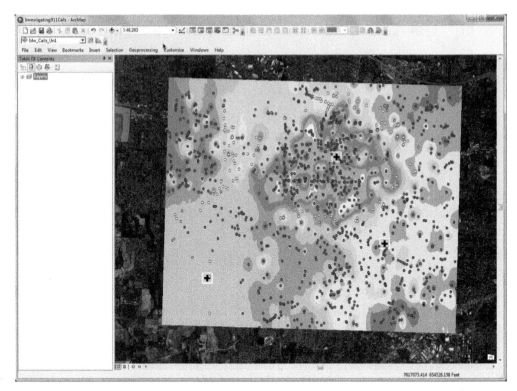

FIGURE 2-3 911 Hot spots (dark ovals) demonstrate how certain calls for emergency service often occur in certain groups or clusters.

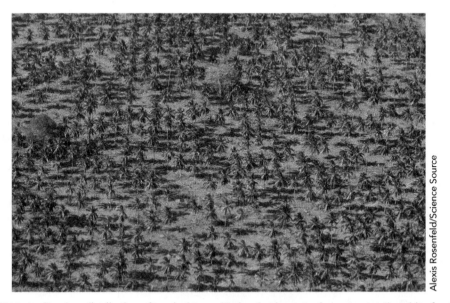

Alexis Rosenfeld/Science Source

FIGURE 2-4 Random distribution of tropical trees. Notice that in spots, there seems to be a bit of order and in others more of a clustering. This clearly demonstrates that random is somewhere between uniform and clustered distributions.

The reasons for the variety of geographic patterns generally have much to do with some underlying process that caused them to occur where and when they did. One can assume that, because there is neither a strong clustering nor a strongly uniform pattern, whatever forces are at work are also random. As before, this relationship between pattern and process is at the heart of the geographic method of identifying patterns and seeking explanation for their nature, causes, and impacts. GIS is all about the discovery, description, explanation, and prediction of these patterns of geographic phenomena using the computer programs built into the GIS software.

ACTIVITY 2-1   CHARACTERIZING DISTRIBUTIONS

In this activity, you will be given an opportunity to demonstrate that you can identify each of the three basic spatial patterns you have just read about. You will also be challenged to select which of the following patterns relates to selected types of activity.

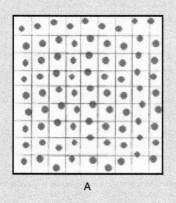
A

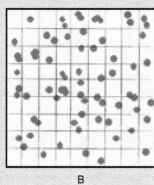
B

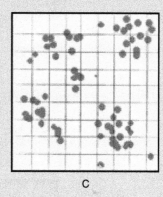
C

1. Which of the distributions above (A, B, or C) would be classified as random? _____

2. Which of the distributions above (A, B, or C) is most likely associated with some form of human activity?

_____

3. Which of the distributions above is most closely associated with what crime mappers would call "hot spots"?

4. Provide a short description of the idea of pattern and process and use figures A, B, and C as examples of linking pattern and process.

_____

_____

_____

_____

5. Why is it important for a GIS practitioner to know and understand the idea of pattern and process?

_____

_____

_____

_____

Further Exploration

Go onto the Internet and use your favorite browser to gather a collection of images of geographic features that display regular (uniform), random, or clustered distributions. See if you can guess what specific processes might have caused these distributions. When you have finished, share these with your classmates. Your instructor might even assign this as a discussion activity.

Geographic Dimensionality

In the previous section, you were introduced to distributions of geographic objects, but there are many types of geographic data that make up these distributions. Geographers group geographic data into four basic types based on their dimensionality: points, lines, areas, and surfaces. Each of these types of data has properties associated with its specific dimensionality (i.e., 1-, 2-, or 3-D). These properties are length, width, height, and volume. The following discussion focuses on the impact of an object's geometry on the GIS analyst's ability to characterize, measure, and compare its features.

Points

Points have zero dimensions. They have no length, no width, and no height. Points can be real objects like individual volcanoes, wild animals, towns, churches, hospitals, supermarkets, oil wells, mines, benchmarks, and a host of other objects we see every day. Points don't always have to represent specific objects in geographic space. Instead, they can represent the locations of events such as earthquakes, crime occurrences, traffic accidents, geotagged locations, chemical spills, fires, disease occurrences, and sales.

Although very simple, point objects tend to confuse people a bit because they can't imagine objects (like supermarkets or towns) as being zero dimensional. Obviously, point objects have dimensions or you couldn't see them, shop in them, or live in them. What is happening is that the geographer has made a conscious decision to think about these features as being so far away that they appear like dots. When you work with your GIS software and you have point objects in your database, you will find that the software recognizes this zero dimensionality because as you zoom in on the point, its size, often represented by a single dot on the computer screen, remains the same. No matter how many times you try to zoom in on it, the dot never gets any bigger (Figure 2-5).

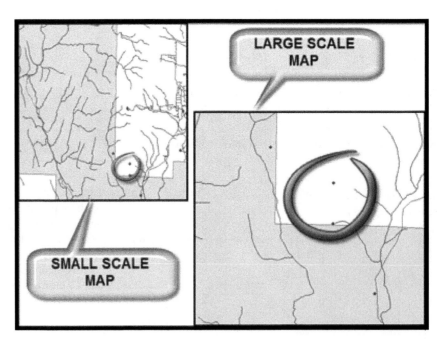

FIGURE 2-5 Points do not get larger when you zoom in with your GIS software because the software recognizes that points are zero dimensional. Notice that after zooming in on the map at left, the points on the zoomed map at right have not increased in size.

Lines

Line or linear objects are conceived of as having only one dimension—length—so they are called one-dimensional (1-D). As with points, lines can be concrete objects such as fences, rail lines, and cliffs, or they may be contrived like state or national boundaries, boundaries between soil types, or flight lines for an airline.

Lines, like points, are defined partially by the scale at which they are observed. A row of houses at one scale, for example, might be considered an area and at a very small scale as a set of points. At yet another scale, the analyst might consider them to have only one dimension and thus will work with them at that level (Figure 2-6). Because linear objects have that one dimension—length—the analyst now has the ability to describe features that were unavailable for point objects.

The GIS analyst now has the option of measuring that length. That ability to measure length gives the analyst considerably more power than when working with points because points cannot be measured. Keep in mind also that lines can be straight or curvy, so you can measure the amount of curviness. They can also form **networks**, such as the U.S. interstate highway system. There are special GIS programs that are specifically designed to analyze aspects of these networks. You'll be introduced to those in Chapter 7. Finally, again as with points, line symbols may take up two dimensions of space, but the objects themselves are thought to be only lines that have no thickness at all.

Areas

Areas, called **polygons** by GIS professionals because that is what their geometry is, are 2-D, that is, they have both a length and a width, both of which can be measured. Common examples of areas might be forests, pastures, industrial parks, or lakes. As before, these must be viewed at a large enough scale to allow the detail you need to view them as areas.

Given that polygons have an additional spatial dimension, they provide GIS professionals like you still more options for description, measurement, and analysis with polygons than with points or lines. You can measure the polygons' longest or shortest axis, their perimeter, and of course their area. And you are not limited to those measurements because you can compare measurements. One common comparison in GIS analysis is the long versus short axis so you have some idea of whether the polygon is long and skinny or rounder. A similar measurement GIS analysts can make is the ratio of the perimeter to the area—again giving a simple measure of the shape of the polygon.

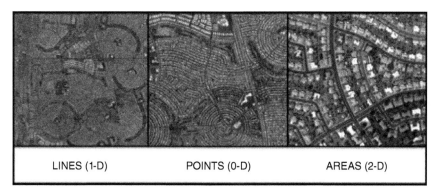

LINES (1-D) POINTS (0-D) AREAS (2-D)

FIGURE 2-6 Three images of Sun City, Arizona at different scales. Notice that, depending on how much you zoom in, the dimensionality of the objects under consideration changes. On the left, the houses look like lines, in the middle, like points, and on the right like areas.

Polygons are very easy to work with in GIS because the earliest forms of GIS were based on the idea of storing, measuring, and comparing different polygons. You might recall from Chapter 1 the idea of co-occurrence of different features such as the relationship between soil polygons and associated vegetation polygons. Polygons are also normally connected (**adjacent**) to other polygons. Such relationships, as you will see in several of the following chapters (especially Chapters 8 and 9), are often very important. Imagine, just as a brief example, that you own a piece of land (a polygon) that just happens to be right next to a polygon that the city wants to turn into a sewage lagoon. Imagine how this might affect the value of your home, not to mention your sense of smell.

Surfaces

Surfaces are 3-D features in geographic space. The most obvious example is the topographic surface composed of an infinite number of possible elevation values spread over an area.

While we usually think about topographic surfaces, there are many more that we encounter every day. Meteorologists, for example, will make maps of temperature or barometric pressure. Population specialists (demographers) might want to see trends in population and will make maps showing a surface based on population numbers. Agriculture specialists might create surface maps of heating degree days. Geologists might make surface maps of the magnetic force emanating from different places. There are actually quite a few nontopographic surfaces from which to choose. Surfaces that have numeric values associated with them are called **statistical surfaces**. The idea of statistical surfaces is an important one because the GIS analyst has lots of statistics (numbers) that can be analyzed to predict trends, to make reasonable guesses at missing values, and to determine the relative steepness of one part of a surface to another. In Chapter 8, you will get a chance to practice analyzing surfaces.

Surfaces come in two major types: discrete and continuous.

Discrete Surfaces

A discrete surface is formed by connecting a series of flat surface segments within which all the values are the same. In short, they are blocky and look a bit like steps rather than having smooth continuous changes (Figure 2-7). The primary characteristic of the discrete surface is that the data do not occur everywhere but only at discrete locations—hence the name.

Discrete surfaces are less familiar to most analysts but occur with surprising frequency. The vast majority of socioeconomic and demographic data that make up, for example, the U.S. Census

DISCRETE SURFACE

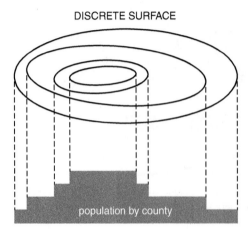

population by county

FIGURE 2-7 Discrete surface represented by population collected by county. Notice that each portion of the surface has a very abrupt change. This indicates that the surface is discrete.

CONTINUOUS SURFACE

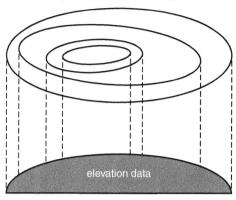

elevation data

FIGURE 2-8 Continuous surface represented by elevation data collected at individual sampled point locations. In this case, the surface is smooth, reflecting the existence of an infinite number of potential sample locations.

Bureau's datasets are by their very nature discrete. These datasets present some limitations in how they can be analyzed and how they are presented. One critical limitation is that because one cannot assume continuous change, one is also not able to predict unknown or missing values (a process called **interpolation**) with the same level of certainty one might expect from continuous data.

Continuous Surfaces

The more familiar surface to most of us is the continuous surface. The primary characteristic of the continuous surface is that the data occur in an infinite number of possible locations such as individual measurements of temperature, elevation, and barometric pressure (Figure 2-8).

Because continuous surfaces have an infinite number of potential data points, they present the necessity of sampling to reduce the number of points. You will learn more about this in Chapter 3. In the meantime, you should be aware that the continuous nature of this type of surface offers some really nice opportunities for analysis, particularly for topographic surfaces. You will encounter this in Chapter 8, but you should be aware of the differences between discrete and continuous surfaces because they are treated differently in both analysis and display.

The reason for these differences might just be the nature of the surface, like buttes and mesas in a desert landscape, or be related to the way the data that comprise these surfaces are actually collected. Examples might include data collected by areas rather than by points such as population by county, total farm income by township, or number of college graduates by city. The important characteristic of discrete surfaces is that the data do not occur everywhere but in only specific areas (i.e., discrete locations).

ARCGIS LESSON 2-1 | GETTING TO KNOW YOUR GEOGRAPHIC FEATURES

In this first ArcGIS Lesson, you are going to use the online version of Environmental Systems Research Institute's (Esri) ArcGIS software to examine some maps from its map gallery and answer a set of basic questions.

1. Use your favorite Internet browser and go to http://www.arcgis.com.

2. Go to the Gallery.

3. Click on all of the maps you see in the Gallery. When you click on each, it will expand. *Note:* When you expand a map, you will see on the left a number of options. On the top, you have three views including one for adding a map legend (right). Below you can get detailed map information, load the map into two GIS products (ArcExplorer and ArcGIS), or even edit or create your own map using ArcGIS online.

(*continued*)

ARCGIS LESSON 2-1 | (CONTINUED)

4. Examine at least 10 maps from the gallery.

5. For the maps you examined, fill in the following table.

Keep this table in mind because you will be referring back to it in a future ArcGIS Lesson exercise.

Map Title	Purpose (theme)	Dominant Feature Dimension (0, 1, 2, or 3)	Secondary Feature Dimension (if present) (0, 1, 2, 3)	If 3-D. Is It Continuous or Discrete?

Data Measurement Levels

When you first think about geographic features such as buildings (Sears Tower), streets (Interstate Highway 10), land use categories (row crops or urban), and even mountain ranges (the Rocky Mountains), it's pretty common to envision them by first naming them. Another important aspect of geographic features for the computer to understand is that the features can be quantitative (have numbers assigned to them) rather than just qualitative (named or described). That geographic features can be represented by numerical data is critical to quantitative analysis inside the GIS software. Not only do these quantities provide more detail about features on the Earth but also they allow the GIS analyst to make quantitative comparisons depending on the nature and measurement levels of the data that are input into the GIS software.

There are five levels of data measurement that modern GIS software is capable of handling. In order of quantitative specificity, the first four are **nominal**, **ordinal**, **interval**, and **ratio**. The last one, **scalar**, is highly variable and can be either very specific or rather general. Take a look at the following detailed descriptions of each.

- **Nominal.** You have already seen the least specific level of data measurement in Activity 2-1, known as *nominal*. Nominal, or named data, are not based on quantitative measures at all but on some classification system the geographer or other spatial scientist devised to separate one category from another. Nominal data are categories or names of data, and as such, cannot be compared directly to one another. Agricultural land, for example, cannot be compared to building types because the two categories are totally unrelated. Keep in mind, however, that there are many GIS operations that rely on nominal data, so don't dismiss them as unimportant.

You'll get many opportunities to experience the analysis and manipulation of nominal data in many of the following chapters.

- **Ordinal.** One way to remember ordinal data is that they are data that are "in order." In other words, they are ranked. Small, medium, and large would be typical ranks in this system, and they might be applied to towns or ranches. Other common examples of ordinal geographic features include county, state, and U.S. and interstate highways. In the latter example, the names themselves imply the rank and therefore the size and/or volume of traffic these highways might be capable of handling. Ranked data can be compared but only by their ranks. This means that any analysis based on ordinal data is limited to ranking procedures.

- **Interval.** The term *interval* itself helps to describe the nature of these kinds of data in that they are divided into small intervals that allow analysts to be able to measure them very carefully. If, for example, you had soil probes with thermometers in them, you could determine the soil temperatures in degrees Fahrenheit or Celsius. Actually, temperature in Fahrenheit or Celsius is a good example of interval data because it demonstrates one property that limits interval data—the ability to make a ratio. For example, if it is 50° F in Nome, Alaska, and 100° F in Miami, Florida, it might be tempting to say that Miami is twice as warm (a ratio) as Nome. But that would be incorrect because there is no absolute starting point for Fahrenheit or Celsius temperatures. While the starting points (e.g., 0° C) might have a reason to exist (the freezing point of water), it is not an absolute starting point, and ratios would be incorrect.

- **Ratio.** The limitation just described of not having an absolute starting point and therefore lacking the ability to make ratios is overcome with ratio-level data. Ratio data are just like interval data except for the ability to make a ratio. For example, if you owned 100 acres of land in Miami and a friend of yours owned 50 acres in Nome, you own twice (a ratio of 2 to 1) as much land as he or she. From an analytical perspective, the ability to make ratios gives you some capabilities for GIS analysis that interval data don't.

- **Scalar.** There are times when the traditional four measurement systems you just encountered either are not known or cannot be measured. The classic situation is when you are trying to describe something like the difficulty level of climbing a hill or mountain. You may know that if a surface is flat, there is no difficulty in your walk, and if there is a vertical smooth cliff, the difficulty level is absolute. This means you may know the lowest and highest levels accurately, but everything in the middle is pretty much a guess based on your experience. GIS allows you to assign these internal values using a level of measurement called *scalar.* The scale is created by you or by the opinion of others with experience. It is used in situations when absolute measurements are not possible but you have some feel for what they should be. Although not exact, they allow some very powerful analysis.

ACTIVITY 2-2   RECOGNIZING DATA MEASUREMENT LEVELS

In this activity, you will practice working with features that exist at different measurement levels from nominal through scalar.

1. Place an X in the column representing the appropriate scale level for each type of data indicated on the left.

2. The last rows have no feature. Fill in at least one example of four of the five measurement levels and place an X in the column representing its appropriate scale.

(*continued*)

ACTIVITY 2-2 (CONTINUED)

Feature	Nominal	Ordinal	Interval	Ratio	Scalar
Mine					
Barometric pressure					
Temperature in Celsius					
Annual salary by county					
Land use type					
Highway types (e.g., state, U.S., Interstate)					
Fire hazard rating					
Travel impedance					
Crop type					
Percentage of college graduates					
Hazardous spill intensity					

3. Thought question: If the GIS software understands all these data measurement levels, why do you need to know them? As you ponder this "why" question, think about the following.

   a. Does the computer decide how map data are manipulated?

   b. What happens if I multiply the values of two maps and one set of numbers represents nominal categories (like land use) and the other represents interval or ratio (e.g., elevation).

   c. Does the GIS software know how to assign numbers to things like difficulty levels or travel friction?

   d. If I use incorrect data measurement levels in my analysis and the results of my mistake cost someone money, is the computer going to get sued?

ARCGIS LESSON 2-2 | RECOGNIZING DATA MEASUREMENT IN GIS DATASETS

In the first ArcGIS lesson, you examined a number of maps from the ArcGIS online catalog. Refer back to the list of maps you selected. The following is a portion of the table with no entries.

Map Title	Purpose (theme)	Dominant Feature Dimension (0, 1, 2, or 3)	Nominal	Ordinal	Interval	Ratio

1. Examine the same 10 maps you did in ArcGIS Lesson 2-1, but pay particular attention to the legend this time.

2. For each of the 10 maps you examined identify, by placing an X in the appropriate box, the level of geographic data measurement represented.

Note: You will notice there are no scalar data present as this is less common and is not often represented in the Gallery.

3. Optional question: Your instructor might want the class to compile a comprehensive list of these maps to share with all the learners.

Map Symbolism

Maps are fundamental to GIS both as data input and as the output from analysis. In creating a map, the cartographer employs a set of lines, colors, and marks collectively called map **symbols**. These symbols allow the map reader to identify geographic features that have been put on the map and to determine their positions visually. Unfortunately, for symbols to be seen, they must by definition take up at least a small portion of space. A consequence of this is that there is often a difference between the dimensionality of the geographic features and that of the symbols themselves. The classic example of this is with point objects that, as you might remember, have zero dimensions. If the symbols for points were truly zero dimensional, they would not be visible. In fact, if line symbols had only the one dimension—length—they would be invisible because lines have to have width (the second dimension) to be big enough to be seen.

Although map symbols are generally 2-D, they must represent both the dimensionality of geographic features (i.e., 1-, 2-, and 3-D) as well as the correct measurement levels—especially nominal, ordinal and interval/ratio. You might be wondering why I combined the last two measurement levels (interval and ratio). This is because the symbols for both interval and ratio are identical. The relationships between data dimensions and data measurement levels are commonly summed up in diagrams like the one in Figure 2-9.

It is important to be able to recognize these symbols because they communicate the relationships between measurement levels and symbol type. It is also important just to be able to recognize what is being represented on the map. Also, keep in mind that if you are trying to convert a map data point—say a mine as shown in the upper left of Figure 2-9—into a digital

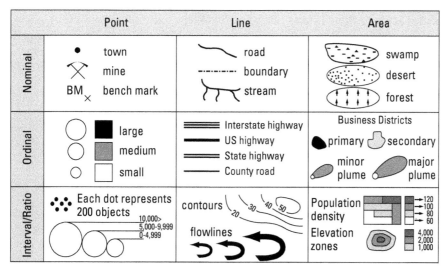

FIGURE 2-9 Diagram showing the relationships among geographic symbols (points, lines, and areas) and their scales of data measurement (nominal, ordinal, and interval/ratio).

equivalent, the symbol takes up space, and that means that the true location of the object is "somewhere" in that space. This means that the very act of converting map data to digital GIS data introduces error. It is also important to notice that there isn't always a one-to-one relationship between the dimensionality of the symbol and that of the geographic feature you are working with. Notice, for example on the bottom center, that lines—contours in this case—can represent surface features. If you haven't studied map reading or cartography, it will take a while for you to get used to these symbols, but don't despair. With just a little bit of practice, you will pick this up.

Geographic Scale

The discussion in the section 'Types of Distributions' regarding dimensionality shows its underlying dependency on one of the most important concepts in GIS—that of map scale. A working definition of scale is "the ratio of the map units to the earth units." Scale can be depicted on the map as a graphic device sometimes called a **bar scale** (Figure 2-10). Bar scales are useful for map output because their size changes in direct proportion to the size that the map is displayed. This is a good thing to remember when producing maps because it is easy to put text on a map that indicates that it is a particular scale and then print the map at an entirely different scale with that text still there. Scales are also depicted with numbers in two primary forms. One of these scale depictions is called the **verbal scale** such as 1" = 36 miles; that allows you to put different measurement units on either side of the equals sign. The other scale depiction is called the **representative fraction** (RF) that uses identical units on either side of the equals sign. The RF doesn't display the units of measure because they cancel out of the equation (e.g., feet/feet). On today's GIS, the most common form of representation during most phases of operation is the RF, but for display, it is also common to use the bar scale to avoid mismatches between real and displayed scales.

The dimension (e.g., 0, 1, and 2) of spatial data is based largely on the scale at which they are either observed or measured. Recall Figure 2-6 that illustrates three views of the same area at different observational levels. The result is that while they are entirely the same geographic features, they can be interpreted as 0-D (points), 1-D (lines), or 2-D (polygons), depending on the scale of which they are observed. This is critical for the GIS analyst to understand because the nature of the analysis will be altered depending on the dimensionality.

As you will see in Chapter 3, scale has much to do with the level of detail that is available for input to the GIS and that directly affects your ability to do GIS analysis. The more detailed the data—that is, the larger the scale of the map—the more information you will have and the more detailed your analysis can be. While the GIS can change the scale of

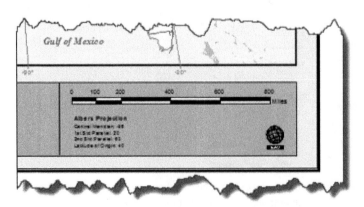

FIGURE 2-10 Bar scale from an ArcGIS map.

representation of the map, it cannot add detailed data to a map layer whose data were collected at a very small scale. So, for example, if you digitized a map of the rivers of the United States from a map at a scale of say 1:1,000,000, you would not be gathering much information on tiny rivers and creeks or on small tributaries of large rivers because they would not appear on the maps that you digitize. As such, you would not be able to analyze those small rivers, creeks, or tributaries even if you changed the scale of the map within the GIS software because the data are simply not there.

In the section 'Map Symbolism,' you learned about map symbols. If scale affects the dimensionality of the objects on the map, it also affects the output of the map itself and how those symbols have been applied to the map. As map scales get smaller (i.e., as you cover more area with a single map), the less detail can be shown on the map you display. To symbolize these smaller scale maps, some features might have to be simplified and others might even be eliminated to prevent clutter on the map document. While the data are retained in the GIS database, they are selectively simplified or eliminated based on rules within the GIS software. This process is complex and has been studied for decades by cartographers, but for most GIS operations, the built-in rules of your GIS will be sufficient to produce adequate map output without much additional manipulation by the user. Still, you should be aware that such simplifications will affect the appearance of the map and might impact the quality of the map reading experience.

ARCGIS LESSON 2-3 | WORKING WITH SCALE IN GIS DATASETS

This ArcGIS lesson is a prepackaged lesson provided by Esri (the ArcGIS people) and designed to allow you to practice with map scale inside a functional GIS. If you are comfortable with the idea of scale, your instructor might suggest you skip this. The lesson is completely self-contained and includes datasets, exercises, and recommended output.

http://edcommunity.esri.com/resources/arclessons/lessons/s/spatial_math_studying_scale4

Projections, Datums, and Reference Systems

The Earth Graticule

One important aspect of GIS is that users will almost always need to know where geographic features are. One uses the power of GIS to locate features, to compare those locations to others, and to make measurements. The most basic reference system that is used to perform these important tasks is called the **graticule** but is more commonly known simply as the **latitude** and **longitude** system of coordinates. Briefly, the earth is divided into two sets of imaginary lines—**parallels** that are lines drawn horizontally and parallel to each other from the equator. Parallels measure latitude in angular degrees from the equator to each pole at a maximum of 90° north and south (Figure 2-10). Other lines called **meridians** are drawn vertically from their start at the **Prime Meridian** that runs through the observatory at Greenwich, England. Meridians are not parallel but converge at the poles. They measure longitude in degrees east and west of the Prime Meridian to a maximum at 180°, a location known as the **International Date Line** because the days are different on either side of that line (Figure 2-11).

Reading locations on the graticule is relatively easy, but carrying a globe around would be cumbersome. Fortunately, virtually all GIS software converts existing flat maps into 3-D coordinates and stores them digitally in the computer. The maps from which that 3-D database was derived, however, were flat. And maps that are produced by the GIS will also be flat. Because the

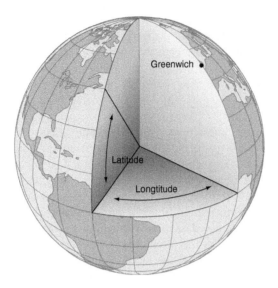

FIGURE 2-11   The Graticule.

earth is not flat, the software must be able to move back and forth between spherical earth coordinates and flat map coordinates. The GIS analyst needs to be aware of the basic process and limitations of converting a sphere (the earth) into a flat map even before any reference system is applied. This process is called *map projection.*

Map Projections

With the exception of globes, all maps, digital map displays, and even 3-D simulations of the earth are essentially 2-D. This means the GIS analyst must somehow convert the earth (3-D) into a flat representation. This is done in two steps. The first step involves changing the scale of the earth into something more manageable by humans. This scaled-down version of the earth called the **reference globe** will provide the basis for all other scale and projection changes.

Once the scale change is accomplished, the projection itself takes place. The mathematics of map projections involves analytical geometry and calculus, but conceptually, it's pretty easy to visualize what is happening. Geographers use the term *projection* because, in effect, they imagine what the earth would be like if a light bulb were placed on the inside of the earth and the light rays spread outward and displayed the results onto one of three families of surface: planar (like a picture of the earth in space), conical, and cylindrical (from left to right on Figure 2-12). Each of these families produces a different map, and each has properties that the GIS analyst needs to know for input, analysis, and output. Keep in mind that when I refer to map properties here, I am also including the GIS datasets that represent the maps as well.

The specific properties that I am referring to have to do with the nature of how the process of creating a 2-D map from a 3-D object like the earth distorts the earth's surface no matter how hard the cartographer might try to avoid this. There are four primary properties of the earth that are affected by map projections, and their distortions will affect the accuracy and the visual appearance of the maps that result from GIS analysis.

1. **Angle.** Some projections attempt to preserve angular relationships for small portions of the map and are sometimes called *conformal* or *orthomorphic* projections. Any two lines on such maps intersect the same as they would on a globe. Such map projections can help preserve overall shape, but the smaller the scale, the less likely this is to be true.

2. **Direction.** Travel navigation requires knowledge of **bearing** (direction on a surface), and some map projections are better at preserving direction than any other property.

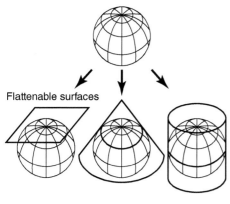

Flattenable surfaces

Flat maps

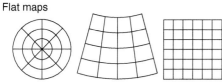

FIGURE 2-12 The general families of map projections—planar (left), conical (center), and cylindrical (right)—based on the surface upon which the map is projected mathematically.

3. **Area.** Among the most useful general-purpose maps are those whose projections preserve area. Such projections, called **equal area** or **equivalent** projections, allow for proper comparison of areas from place to place and for the proper sizing of continents on small scale maps.

4. **Distance.** Maps that attempt to preserve distance are actually controlling the scale of the map. If the scale changes from place to place on the map because of the distortion caused by projecting a sphere onto a flat surface, the distance is also going to change from place to place. Such map projections are useful for navigation.

There are hundreds of map projects, and each has its own ability to preserve or not to preserve these four properties. Table 2-1 is a short list of some common map projections, the family to which each belongs, and the properties it preserves. Notice that some of these projections preserve a selected property entirely, and others preserve it only partially.

Table 2-1 A short list of common map projections, their families, and the primary properties they preserve. Some projections, such as the Robinson projection, are meant primarily for visual display and are therefore often a compromise in that the projection does not preserve any property in its entirety. http://egsc.usgs.gov/isb//pubs/MapProjections/projections.html

		Summary of Projection Properties			
Projection	Type	Conformal	Equal area	Equidistant	True direction
Globe	Sphere	★	★	★	★
Mercator	Cylindrical	★			x
Transverse Mercator	Cylindrical	★			
Oblique Mercator	Cylindrical	★			
Space Oblique Mercator	Cylindrical	★			
Miller Cylindrical	Cylindrical				
Robinson	Pseudo-Cylindrical				

(continued)

Table 2-1 (*continued*)

| Projection | Type | Summary of Projection Properties | | | |
		Conformal	Equal area	Equidistant	True direction
Sinusoidal Equal Area	Pseudo-Cylindrical		*	x	
Orthographic	Azimuthal				x
Stereographic	Azimuthal	*			x
Gnomonic	Azimuthal				x
Azimuthal Equidistant	Azimuthal			x	x
Lambert Azimuthal Equal Area	Azimuthal		*		x
Albers Equal Area Conic	Conic		*		
Lambert Conformal Conic	Conic	*			x
Equidistant Conic	Conic			x	
Polyonic	Conic			x	
Bipolar Oblique Conic Conformal	Conic	*			

Key: * = Yes / x = Partly.

Because maps come in different projections and because you will be working with multiple maps in the same GIS database, you will need to know the specific projection for each. As you input maps into a GIS, the software will ask you for the map projection so that it can keep a record of that information for later use. Many GIS analyses involve the interaction of more than one map, and the software will be able to transform from one projection to another but only if the correct information is included for each map. Ultimately, when you have finished with any analysis, you will also frequently produce maps as the output. When you produce a map, you will also need to keep the properties you want to preserve based on the intended use of your map. Table 2-2 provides a brief list of some of the more common uses of different types of maps based on their projections.

Table 2-2 A selection of major map projections, their families, and some uses that they are compatible with. You will notice that the preserved properties of these projections are included in Table 2-1. http://egsc.usgs.gov/isb//pubs/MapProjections/projections.html

| Projection | Type | Summary of Projection General Use | | | | |
		Topographic Maps	Geological Maps	Thematic Maps	Presentations	Navigation
Globe	Sphere			*	*	
Mercator	Cylindrical	*	*			*
Transverse Mercator	Cylindrical	*	*			
Oblique Mercator	Cylindrical	*				
Space Oblique Mercator	Cylindrical	*				
Miller Cylindrical	Cylindrical			*		
Robinson	Pseudo-Cylindrical			*	*	

Sinusoidal Equal Area	Pseudo-Cylindrical			*		
Orthographic	Azimuthal					
Stereographic	Azimuthal	*	*			*
Gnomonic	Azimuthal				*	*
Azimuthal Equidistant	Azimuthal	*				
Lambert Azimuthal Equal Area	Azimuthal			*	*	
Albers Equal Area Conic	Conic			*	*	
Lambert Conformal Conic	Conic	*	*		*	*
Equidistant Conic	Conic					
Polyonic	Conic	*				
Bipolar Oblique Conic Conformal	Conic		*			

Key: * = Yes / x = Partly.

Datums

Earth **datums** define the size and shape of the earth and the origin and orientation of the coordinate systems used to map it. These datums are based on a scaled model of the earth sometimes called a **reference ellipsoid**. That model is not a perfect sphere, however, as the earth itself is also not perfectly spherical. Rather than dealing with all the minor topographic features when trying to prepare maps, cartographers use a generalized geometric model of the earth that is based on a flatness ratio of the length of the polar axis compared to that of equatorial axis. These ratios are based on a number of different measures and approximations performed by different nations and different surveyors at different times. Because all maps are based on one of these datums, all measurements of the earth depend on which datums are used. As with map projections, you need to know which datums each map uses if the maps are going to be compared in GIS analysis. Most GIS software will ask which datums are used for each map as it is put into the GIS in the first place.

While there are literally hundreds of datums around the world, the two most often used in North America are the North American Datum 1927 (NAD 27) and the North American Datum 1983 (NAD 83). NAD 27 is based on an ellipsoid created in 1866 by manually surveying the North American continent. With the advent of satellite remote sensing and computer technology, the newer NAD 83 is based on an ellipsoid called GRS 80 that is more of a global than regionally derived approximation of the earth's overall geometry. Mixing datums when trying to match GIS datasets will often present some unwanted results.

Reference Systems

During the projection process, it is necessary to transfer the graticule coordinates to a 2-D surface with as little distortion as possible. The resulting flat reference system is linked to the projection within which it was converted. There are a great many reference systems, each designed for particular portions of the earth and for different reasons. Perhaps the three most relevant to students in the United States are the Universal Transverse Mercator (UTM) system, the State Plane Coordinate (SPC) system, and the U.S. Public Land Survey (USPLS) system.

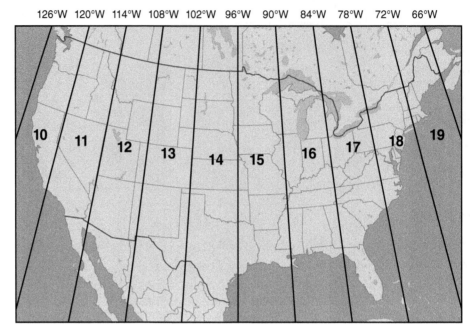

126°W 120°W 114°W 108°W 102°W 96°W 90°W 84°W 78°W 72°W 66°W

FIGURE 2-13 Simplified zones of the UTM coordinate system in the United States, each of which is 6° in longitude wide.

Among the more commonly used coordinate systems is the UTM system that divides the earth between 80° S Latitude and 84° N Latitude into 60 zones (Figure 2-13), each of which is 6° of longitude in width. Coordinates are measured based first on the number of the zone, whether it is north or south of the equator, and then in meters north (northings) and east (eastings). The point of origin of each UTM zone is the intersection of the equator and the zone's central meridian. In order to avoid the need for negative numbers, the central meridian of each zone is given a "false easting" value of 500,000 meters. In the northern hemisphere, the northings are measured from the equator while in the southern hemisphere there is a false origin at 10,000,000 meters south of the equator at 80° S Latitude.

A relatively more accurate system in the United States is the State Plane Coordinate System begun in the 1930s and linked to the National Geodetic Survey. It consists of 124 geographic zones. Each U.S. state contains one or more of these zones, and each zone uses its own projection. Like the UTM system, measurements are always positive in the north and east directions. Although for short distances the SPC is very accurate, this accuracy drops off outside of individual zones. This means the SPC is not useful for regional or national mapping.

An older land records system, the origins of which date back to the Land Ordinance of 1785, called the U.S. Public Land Survey was developed for large portions of the United States. This system uses a set of lines that divide the land holdings into units of area 1/640 of 1 square mile called **acres**. Land is organized into 36 of these square mile units (sections) collectively called **townships**. Because each section has a known area of 640 acres, it can be further divided into halves and quarters of sections, the size of which is then easily determined by multiplying the portion of land (e.g., ¼ section) by the number of acres in a section (640), making a ¼ section 160 acres. Although not linked to any particular map projection or reference globe, it does use a series of lines that allow one to identify location. Baselines (drawn east–west) allow you to measure township lines north and south, and Principal Meridians (drawn north–south) allow the measurement of range lines east and west (Figure 2-14).

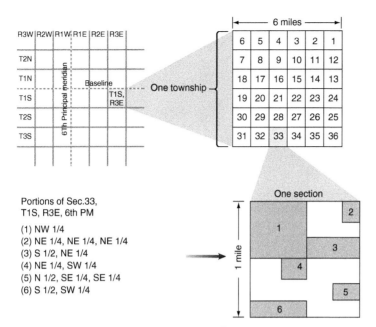

FIGURE 2-14 The U.S. Public Land Survey. Notice that the sections are numbered from upper right in a back-and-forth pattern and ends at the bottom right. Also observe how the portions of section 33 are noted on the bottom left of the figure.

ACTIVITY 2-3 BASICS OF PROJECTIONS, REFERENCE SYSTEMS, AND DATUMS

This activity allows you to examine projections, reference systems, and, to a lesser degree, datums.

1. Answer the following questions.

   a. Why do GIS people need to know about map projections?

   _____

   _____

   _____

   _____

   b. What family does the Mercator projection fall into?

   (1) Mollweide? _____

   http://en.wikipedia.org/wiki/Mollweide_projection

   (2) Sinusoidal? _____

   http://en.wikipedia.org/wiki/Sinusoidal_projection

   c. Describe the changes in circles for . . .

   (1) Mercator projection

   _____

   _____

   (2) Mollweide projection

   _____

   _____

   _____

   (3) Sinusoidal projection

   _____

   _____

   _____

2. Now go to http://www.digital-topo-maps.com/mytopo/topo.jpg and answer the following questions.

   a. What projection does this map use? _____

   b. What do those big red numbers indicate? _____

   c. Based on your answer in part b, what is the name of the land subdivision system? _____

   d. If you look at the square in the middle numbered 10 and you owned ¼ of ¼ of that land, how many acres would you have? _____

   e. What datums are used in this map? _____

(continued)

ACTIVITY 2-3   (CONTINUED)

    f. What are the numbers in blue around the map? _____

    _____

3. Thought question: Why is it important for you to be aware of projections, datums, and reference systems? Doesn't the GIS handle all this for you? _____

ARCGIS LESSON 2-4 | **USING REFERENCE SYSTEMS, PROJECTIONS, AND DATUMS**

Try these three lessons prepared by ESRI (Environmental Systems Research Institute). They will give you a chance to exercise your skills with projections and coordinate systems.

Spatial Math, The Geometry of Map Projections http://edcommunity. esri.com/resources/arclessons/lessons/s/spatial_math_the_ geometry_of

Exploring Map Projections http://edcommunity.esri.com/resources/ arclessons/lessons/s/spatial_math_exploring_meas3

Spatial Math, The Land of Cartesia: Coordinates http://edcommunity. esri.com/resources/arclessons/lessons/s/spatial_math_the_ land_of_cart

---

Raster versus Vector (Basic GIS Data Models)

Prior to creating a GIS database, the GIS professional must first decide on how the geography is going to be represented. This representation has profound impacts not only on how the computer stores the data but also on how they are retrieved and, most importantly, on how they can be manipulated for analysis. Geographers familiar with computer technology refer to this as a **data model** because it is a conceptual model of how the data are going to be characterized. Two basic ways for this have become accepted in the industry. The first is called the **vector** data model that envisions geographic features as a set of points, lines, and polygons much like any drawing you might normally make by hand. The second data model simplifies the geography into sets of squares called **grid cells**. Modern GIS software is normally quite able to work with either of these data models, but it is useful to look at the similarities and differences both in structure and in functionality.

Vector Data Model

The vector model conceives of geographic space as being represented inside the computer as a series of points, lines, and polygons (Figure 2-15) whose positions are based on X, Y, and, on occasion, Z coordinates.

FIGURE 2-15 Basic vector graphic data representation. The structure shows points as individual coordinate pairs, lines as groups of two coordinate pairs, and areas as connected lines having identical beginning and ending coordinates.

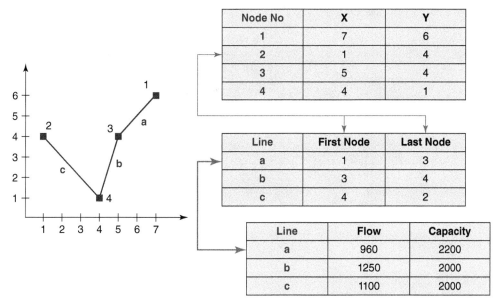

FIGURE 2-16   A basic GIS database with the graphics on the left and the descriptive data stored in relational database management system tables.

Vector systems produce maps that are relatively realistic and require less storage than their counterparts. Most vector systems require a database management system to store the descriptive data while the graphics reside in separate files (Figure 2-16). Much of the analytical work is done using the database management system rather than the graphics directly. There is much to know about the different vector systems, and they will be explained as you begin working with them.

Raster Data Model

The raster model envisions geographic space as a set of square units (grid cells) whose positions are based on column and row positions (Figure 2-17). This model is pretty easy to imagine in that

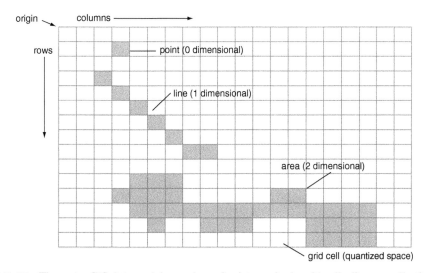

FIGURE 2-17   The raster GIS data model conceives of points as single grid cells, lines as collections of grid cells, and areas as groups of grid cells.

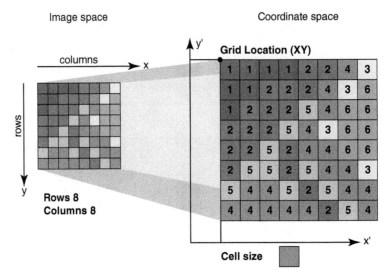

FIGURE 2-18 Typical configuration of a raster GIS where numbers in the grid cells indicate the category or other data represented by each type.

points are represented by single grid cells, lines by long strings of grid cells, and areas by groups of grid cells. In some simple raster GIS software, the grid cells contain numbers that represent the categories or other data for each layer in the database (Figure 2-18). More sophisticated systems also have the ability to store grid cell data as a set of tables so that fewer layers are needed to store the data.

There are advantages and disadvantages for both vector and raster GIS. Vector will produce more cartographic-like output than raster will, making vector the system of choice for display. Because large areas of the earth can be represented as single polygons, the vector model takes up less storage than raster. Raster, on the other hand, has the advantages of speed, modeling power, and ability to easily incorporate remotely sensed data that occur in a similar format. The modeling power of the raster model comes from its ability to store each grid cell as a number, making each part of the map easy to get and allowing for some mathematical manipulation. Although both vector and raster models have a means to store and manipulate surfaces, the raster model is much more flexible. In fact, most surface modeling capabilities in systems with both raster and vector transfer most surface modeling ability to the raster portion of the GIS.

ACTIVITY 2-4   DATA MODELS (RASTER VERSUS VECTOR)

In this activity, you will get a chance to answer some questions about the two primary data models used in GIS.

1. Define the vector data model and provide a labeled diagram.

2. Define the raster data model and provide a labeled diagram of it.

3. Provide at least one advantage of vector over raster and one advantage of raster over vector.

4. Which of the two data models (raster or vector) would be more spatially accurate?

ADDITIONAL READING AND RESOURCES

Flacke, Werner, and Birgit Kraus. *Working with Projections and Datum Transformations in ArcGIS: Theory and Practical Examples, Cartography and Geographic Information Science*, 33(3), 233–37.

PERTINENT WEBSITES

Data Measurement (wiki) http://en.wikipedia.org/wiki/Level_of_measurement

Geodetic Datum Overview http://www.colorado.edu/geography/gcraft/notes/datum/datum.html

Longitude and Latitude Movie http://video.about.com/geography/Latitude-and-Longitude.htm

Map Projections Primer http://www.colorado.edu/geography/gcraft/notes/mapproj/mapproj_f.html

Map Scale Refresher http://egsc.usgs.gov/isb/pubs/factsheets/fs01502.html

Raster versus Vector http://gislounge.com/geodatabases-explored-vector-and-raster-data/

KEY TERMS

acre: A unit of area measure in the U.S. Public Land Survey System equivalent to 1/640 of a square mile.

adjacent: A spatial relationship in which two objects share a common boundary.

bar scale: A method of displaying a map scale using a graphic device that shows how long a portion of the map would be in real earth dimensions.

bearing: Compass direction from a starting point, typically measured in degrees.

clusters: Those portions of a clustered distribution where objects occur in proximity to each other.

data model: In GIS, a computer-based mathematical construct by which geographic data or surfaces can be represented so the computer can use it.

datums: The reference specifications of a measurement system, usually a system of coordinate positions on the earth. Each datum's specific reference ellipsoid is unique to it.

distributions: Collections of distributed phenomena each of which exhibits its own pattern based on underlying processes.

equal area: A form of map projection, sometimes called equivalent, in which an area is the primary property that the projection process attempts to retain.

equivalent: See *equal area*.

geodata: Data whose locations are linked to positions on the earth.

graticule: The name for the spherical grid system—latitude and longitude—that characterizes the earth.

grid cells: Uniform, discrete portions of the earth used in the raster data model. Typical grid cells assume a flat earth and are square but neither of these is an absolute requirement.

hot spots: Portions of a surface that are identified as having high concentrations of phenomena. Centers of high crime are classic examples of hot spots.

International Date Line: An imaginary line of longitude located 180° east or west of the Prime Meridian and in which the time zone east of the International Date Line is twelve hours ahead of Greenwich mean time and the time zone west of the International Date Line is twelve hours behind Greenwich mean time.

interpolation: In surface modeling, the set of mathematical procedures used to predict missing surface values by comparing values on either side of the missing value's point location.

interval: Level of data measurement in which values can be divided into small portions but for which the starting point for measurement is arbitrary.

latitude: Angular distance measured in degrees north and south of the equator. The imaginary lines that represent these measurements on the graticule are called *parallels*.

longitude: Angular distance measured in degrees east and west of the Prime Meridian which runs through Greenwich, England. The imaginary lines that represent these measurements on the graticule are called *meridians*.

meridians: See *longitude*.

networks: Interconnected sets of points and lines that represent possible routes from one location to another.

nominal: A data measurement level that is noncomparative and is based on names or other non-numeric classifications.

ordinal: A data measurement level that consists of ranks of data rather than precise measurements.

parallels: See *longitude*.

persistent: Data that exist for extended periods of time.

polygons: Computer graphic representations of areas consisting of line segments made up of points and where the beginning and ending points occur at the same place.

Prime Meridian: Starting meridian for measuring longitude east and west. The Prime Meridian runs through Greenwich, England.

ratio: A data measurement level that divides data into small intervals, has absolute rather than arbitrary starting points, and allows for the comparison of data through ratios.

reference ellipsoid: An ellipsoid (geometric approximation) used by an associated geodetic reference system or geodetic datum.

reference globe: A hypothetical globe with a standard scale against which map projection properties are compared.

regular: A form of distribution of geographic objects in which the measured distance from one object to another is identical to all other such measurements.

representative fraction: A form of scale notation that is written as a ratio in which the units of measurement on either side of the equal sign is identical and therefore not included in the notation.

scalar: A level of data measurement that is based on a scale relevant to the individual situation or circumstances for the data collector.

spatial: Having to do with or existing within space.

statistical surface: Ordinal, interval, or ratio data represented as a surface in which the height of each area is proportional to a numerical value.

symbols: Markings on a map that represent specific geographic features.

vector: A data model in which geographic space is represented by a series of points, lines, and polygons.

verbal scale: A type of scale representation in which the relationship between map and earth distance is represented by words.

CHAPTER 3

Creating and Editing GIS Data

LEARNING OBJECTIVES

Here is the content you will learn in this chapter:

1. The data formats supported by ArcGIS and where to locate the tools for conversion and interoperability.

2. How to create new features using feature templates.

3. How to create new point, line, and polygon features.

4. How to create new features from existing features.

5. How to create and edit annotations and dimensions.

6. How to edit existing features in a GIS database.

7. How to digitize features.

8. How to edit attributes.

9. How to edit topology.

10. The principles and practice of spatial adjustment.

BEHAVIORAL INDICATORS

When you are finished with this chapter, you will be able to:

1. Identify at least three major data types commonly used in ArcGIS software and five that are not part of the ArcGIS software but are used by it. Identify where in ArcGIS you will find the conversion and interoperability tools.

2. Explain what a feature template is and how to use it to create new features for your project.

3. Provide a description with graphic sketches of at least two different methods used to create point features, line features, and area features.

4. Provide a description of at least two different methods used to create features from existing features in your database.

5. Define *annotations* and *dimension* and describe at least two methods for creating them in ArcGIS.

6. Describe the editing process in ArcGIS including a description of the editing tool.

7. Describe the proper steps in the process of digitizing a map.

8. Explain the process of editing attributes (descriptors) in ArcGIS.

9. Define *topology* and explain the process and importance of editing map topology to the overall quality of your database.

10. Describe the purpose of spatial adjustment and the general steps in performing spatial adjustment.

Chapter Overview

This chapter introduces you to the types of data that can be included in your ArcGIS project and to the basics of creating and editing the **entities** (graphical objects) and **attributes** (descriptive information) inside ArcGIS. You will first be introduced to the many types of data that are compatible with ArcGIS so you will understand its ability to incorporate them. Next you will learn how to create GIS datasets based on templates and how to create them independently. You will encounter different methods of input and will get hands-on experience editing datasets. Besides editing the features and attributes, you will also learn the basics of editing **topology**, the explicit geometry and mathematical rules that govern how the features are correctly organized in the database. Finally, you will learn the basic principles and practice of spatial adjustment, which you will need to match adjacent map sheets and line up features in multiple maps. All of the principles contained in this chapter are critical to producing a good, clean, organized database that you will need if any analysis is to be effective.

Geographic Data Formats

A critical activity of GIS is the input, storage, and editing of appropriate data for the projects that you will analyze. Not only is it critical, but also it consumes a major portion of the time that GIS technicians spend interacting with the software. The following paragraphs provide you with necessary skills needed to create, store, and edit databases, but before you learn that, you will learn that GIS data come in many formats and types and being aware of this makes you a better-informed and more valuable GIS practitioner.

ArcGIS software, like other similar packages has its own "native" data formats—data that the software was originally designed to store, manage, and manipulate. Even native data formats change, however. For example, one of the premier data formats specifically created for Esri's software, originally called ArcINFO©, was the **coverage** data format. This layered data format incorporated the use of topology that is the explicit recording of the entity data geometry that enables you to make sure the data you are working with are accurate. Later, ArcGIS© adopted a new data format called the Geodatabase, which is now the standard for the latest versions of their software. Another major Esri format is the **shapefile** that does not include topology but is easy to work with and very flexible.

There are many commercial GIS and Remote Sensing software packages that use their own data formats. Examples include ERDAS .IMG, .GIS, .LAN, .RAW, and ER Mapper's .ERS and .ECW formats. Additionally, many organizations such as the U.S. Bureau of the Census and the U.S. Geological Survey (USGA) have their own data formats. With all these formats, it is a bit hard to keep track of them all. Fortunately, the folks at Esri have done the legwork for you. In the ArcToolbox menu of ArcGIS, there are two sets of tools for working with all these formats: the conversion tools and the data interoperability tools (Figure 3-1). The conversion tools allow one to convert to and from Xcel, GPS (Global Positioning System), **KML** (Keyhole Markup Language), Raster (grid-cells), **WFS** (Web Feature Service), JSON (JavaScript Object Notation), different metadata types (data about data), CAD (computer-aided design, commonly AutoCAD **.dxf**), Collaborative Design Activity (**COLLADA**), Coverages, dBase databases, **geodatabase**, and shapefiles. This is an impressive set of tools for changing data formats. The data interoperability tools use Safe Software's FME technology to extract, transform, and load a wide range of spatial data

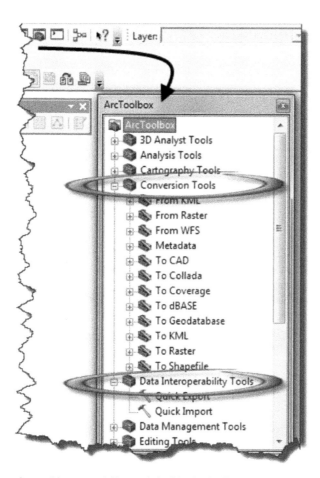

FIGURE 3-1 Conversion and interoperability tools inside the ArcToolbox pulldown menu.

formats. The interoperability tools are simple to use and include Quick Export to convert feature classes or feature layers (the common types of data in ArcGIS) into any format supported by ArcGIS and Quick Import which converts non-ArcGIS data into ArcGIS native formats.

ACTIVITY 3-1   GIS DATA FORMATS

In this activity, you will have a chance to explore some of the data formats you might encounter and to discuss the toolsets you have available in ArcGIS for converting, importing, and exporting them.

1. Describe how you would navigate to the tools inside ArcGIS that employ the FME toolkit to export and import spatial datasets from and to ArcGIS format.

2. Describe how you would navigate to the tools inside ArcGIS to convert different spatial data formats.

3. Perform a WebQuest for the following data formats and sources. Provide the link to the wiki describing each. Provide a brief description of each format.

a. KML (Keyhole Markup Language)

b. WFS (Web Feature Service)

c. JSON (JavaScript Object Notation)

d. DXF (Digital Exchange Format)

e. COLLADA (Collaborative Design Activity)

f. dBase

g. ArcGIS (especially look at geodatabase, coverage, and shapefile)

4. Now go to the website http://en.wikipedia.org/wiki/GIS_file_formats. Briefly describe your reaction to what you find there with emphasis on what you will need to know as a GIS professional.

ARCGIS LESSON 3-1 | EXAMINING DATA TYPES

This lesson will acquaint you with the use of the conversion toolkit, one of the two kits used to convert data formats in ArcGIS. A very common form of conversion is to convert ArcGIS layers or maps to the Google KML format that many non-GIS people can have access to. This lesson gives you a chance to see how this tool works.

1. Migrate to representations, Lesson 1 (C:\\arcgis\tutorial\representations).

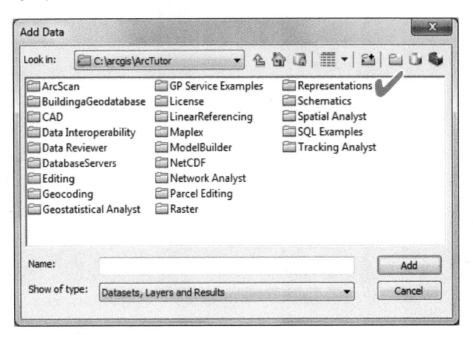

2. Open up the RoadP layer.lyr.

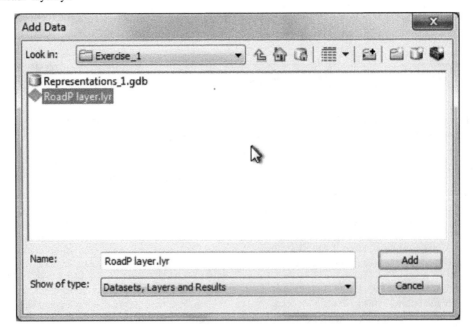

3. A map showing all the cul de sacs in this portion of the Austin, Texas, area is displayed as a series of points, each representing an individual point feature.

4. Open the toolbox (circle) and scroll down to the Conversion Tools and then to "Layer to KML". You are going to convert a layer to a KML file.

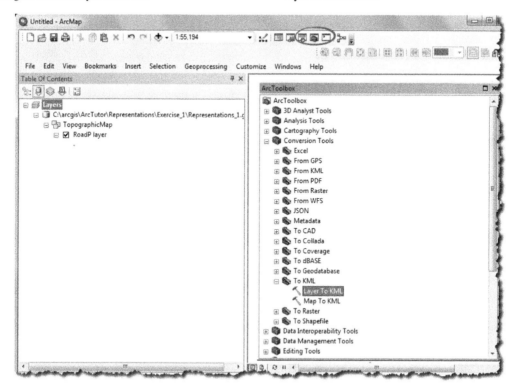

5. Double click on the tool. The following menu opens up:

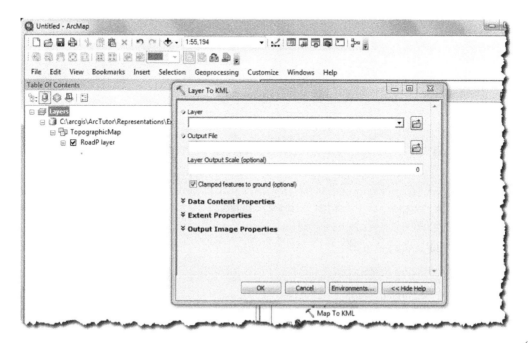

(*continued*)

6. The dropbox opens with three blank spaces. Clicking on each of these provides additional information to help you understand the needed information.

7. In the first space click on the small folder icon and select the RoadP layer.

8. In the second space, define a location for the new layer (where on your computer you are going to put the new layer). You can change the name as well. Remember where you put this file as you will need it again.

9. In space three, the software asks you to select a scale. This is up to you, but one easy way is to approximate the scale of the map in ArcGIS. A scale 50,000 (the denominator for 1:50,000) should work just fine. A dialogue will show that the process is underway, and then a small panel will appear on the bottom right. Clicking on that panel will provide a summary of the process.

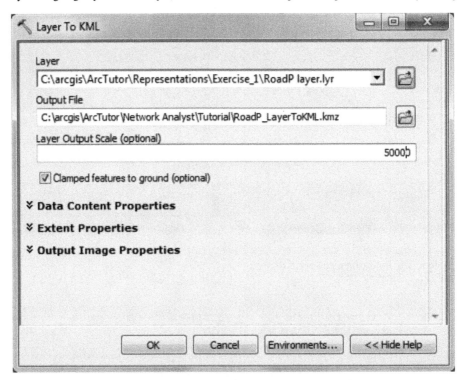

Creating New Features with Template Layouts

In ArcGIS, all data are input from what are called Feature Templates. These templates contain all the necessary information the program needs to create the features in a database including where the layers are stored, what attributes created the feature, and the default tool that is used to create the feature. Also included are

- Feature name

- Feature description

- Features tags for searching

- If templates are not present when you begin editing, ArcGIS creates them automatically for each layer in the current workspace in which you are working. These are saved in the map document (.mxd) and the layer file (.lyr).

One really cool feature of ArcGIS is that you are not limited to creating data on your own. The good people at Esri have created a number of layout templates that allow you to use their existing map projects, datums, and reference structure. Not only can you instantly open up the existing maps from these templates, but also you can add your own data as long as they are located within the geographic boundaries of the regional template you select. This approach is simple, efficient, and straightforward. Its major limitations involve requiring you to begin with fairly large portions of the earth with which to work, but you can edit that later. There might also be a considerable difference in what entities the template contains and those that you need for your project. You can, of course, turn some of this off, add your own, and delete what is not needed. Work through ArcGIS Lesson 3-2 to get a feel for how this works. Take a look at the questions in Activity 3-2 before you begin your ArcGIS Lesson and consider them as you do the exercise.

ACTIVITY 3-2 | **CREATING NEW FEATURES WITH TEMPLATE LAYOUTS**

This short activity reviews what you know about creating point, line, and area features.

Answer the following brief questions before you move to ArcGIS Lesson 3-2.

1. What is the basic advantage of using template layouts to begin your GIS project?

2. What do you think might be some possible negative consequences of using these layouts?

ARCGIS LESSON 3-2 | CREATING NEW FEATURES WITH TEMPLATE LAYOUTS

The purpose of this exercise is to give you an opportunity to learn how to use the template feature of ArcGIS to create maps. This is a rapid way to get started with mapping because much of the basic map infrastructure you need is already there for you.

1. Turn on ArcMap.

2. When the software window opens, place your cursor over My Templates.

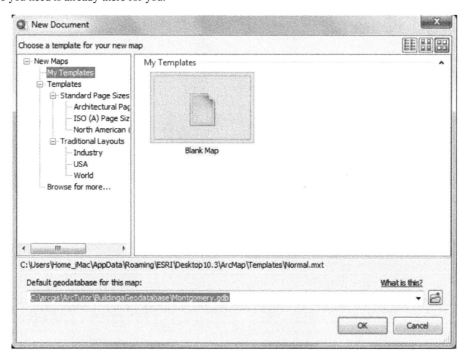

(continued)

ARCGIS LESSON 3-2 | (CONTINUED)

3. Now select World and from that menu, select Australasia.

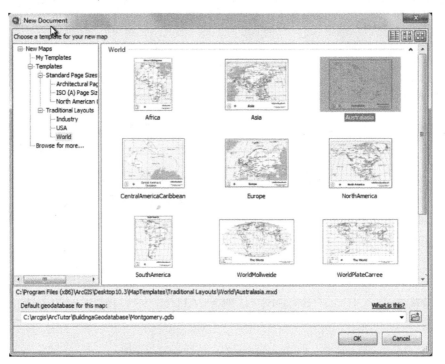

4. Notice the thumbnail that is highlighted. Click on that thumbnail to pull up the map.

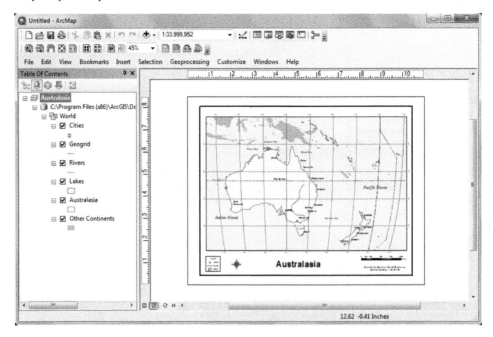

5. Success. Now, select a different region and perform the same steps that you did before.

6. Make a screen shot and turn it in to your instructor.

Creating Point, Line, and Polygon Features

It is quite common to need to add features to a map layer. For example, if you have a roads layer dating from 2010 but new roads now exist, you will need to add the new roads to your map. If new fire hydrants have been added to your town's water network, you will need to add those to your map as well. Consider that each object you insert into your database needs to share the same attributes as its companion objects. Because of this, you can't just drop a point on your map and expect the software to recognize it as a fire hydrant or draw a line and expect your software to understand that the new line represents a road. It would be a really great feature if you could begin with a template that has the basic information present to start with. Enter the ArcGIS templates discussed previously.

ArcGIS allows you to create your own point, line, and area features right on the screen. Before you do this, it is important to understand that points, lines, and polygons by themselves aren't really map features but just graphics. You need to tell the software what the coordinate system, projection, scale, and datum are before the computer will understand that the features you are creating represent real places on the surface of the earth. That's where those feature templates you learned about in the last section come in. Before you begin working on these, it is important to note that each layer can have different templates associated with it. So, for example, if you had a water distribution net, you could have a different template for each of the different fittings such as couplings, taps, risers, and so on. This allows you complete control over your ability to add new connectors of a particular type at will. To create a new tap, for example, you would simply click the tap template, and you can create a new tab by dragging the point object wherever you want (Figure 3-2). What this does is carry all the default attributes of that object along so you don't have to recreate them each time you make a new object. You will have to create any additional attributes you need. These templates work the same way for point, line, and polygon features.

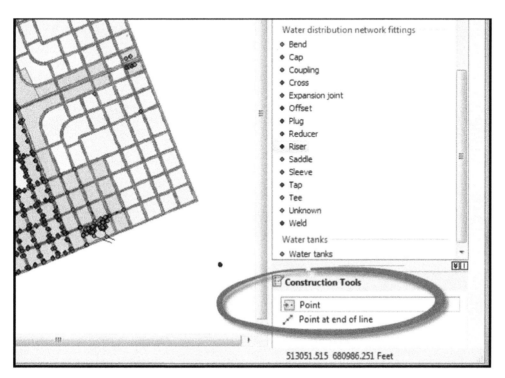

FIGURE 3-2   The construction tools for creating a new point object.

You will get a chance to see this in action in ArcGIS Lesson 3-3. While you work through this brief lesson, try to consider how difficult it would be to have to go through and add all the attributes for each new object manually. Consider also why you might need to add objects to your database in the first place.

| ACTIVITY 3-3 | CREATING NEW POINT, LINE, AND AREA FEATURES |

This short activity reviews what you know about creating point, line, and area features.

Before you begin your ArcGIS Lesson, carefully consider the following questions.

1. Under what circumstances might you find it necessary to add features to a database?

2. What is the basic advantage of using ArcGIS templates for creating point, line, and area features?

ARCGIS LESSON 3-3 | CREATING FEATURES

This short lesson allows you to demonstrate that you can create point, line, and area features using the ArcGIS. A common way to do this is to use aerial photography; in this case, the aerial photography is from Zion National Park. It is a digital orthophoto quarter quad (DOQQ), meaning it is planimetrically corrected and covers one-quarter of a USGS quadrangle map area.

Part I

1. Begin by clicking on the Open Existing Layer button.

2. Migrate to the Editing folder and click on Exercise1 and select Open. An aerial photo of part of Zion National Park appears.

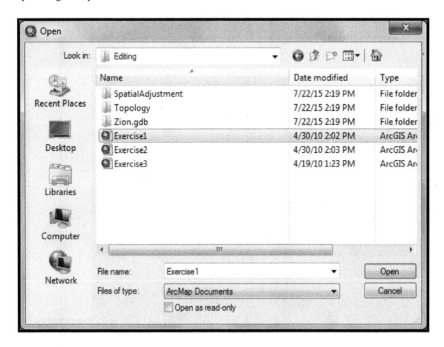

3. If you are prompted to enable hardware acceleration, be careful before clicking Yes. This works on a PC, but users employing a virtual machine inside a Mac environment have had issues with it.

4. Click on the Bookmarks pulldown menu and select Visitor center. *Note:* You can make a bookmark for any view of a map using ArcGIS. Esri has already done this when it prepared the tutorial data.

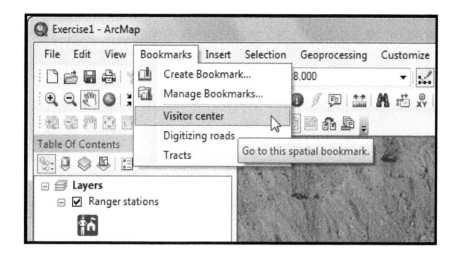

5. Click on the Editor Toolbar button on the standard toolbar.

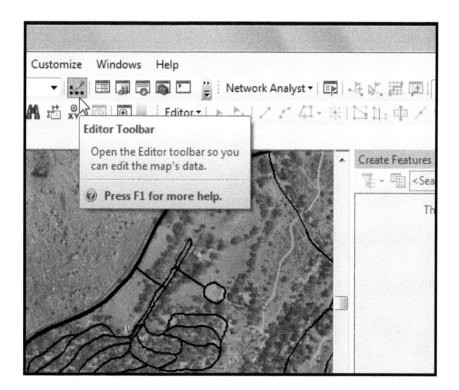

(*continued*)

ARCGIS LESSON 3-3 | (CONTINUED)

6. Click the Editor menu on the Editor Toolbar and select Start Editing.

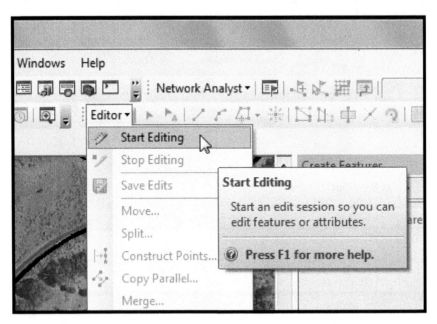

7. The Create Features menu opens up to the right of your map. Click the Ranger stations point feature template. This triggers the editing environment to create point features in the Ranger stations layer.

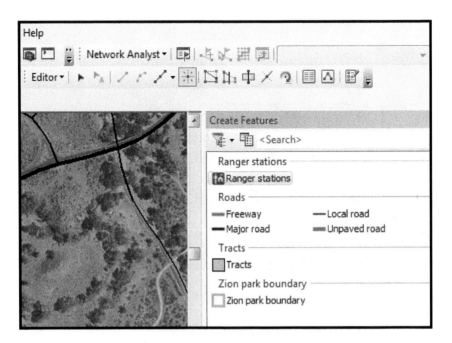

8. Click the Point tool on the bottom of the Create Features window.

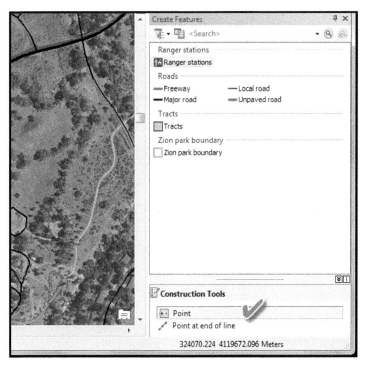

9. On the aerial image, click the map to place a point directly over the visitor center building in the center of the map. Notice the bright blue symbol indicating that the object you just created is also selected (a default condition in ArcGIS).

10. Now click the Attributes window on the main Editor bar or in the Create Features window so you can begin editing the attributes associated with the point you just created for the Visitor Center. Click inside the Location property box that is currently blank and type Visitor Center. Note how the value appears above in the attributes table as well.

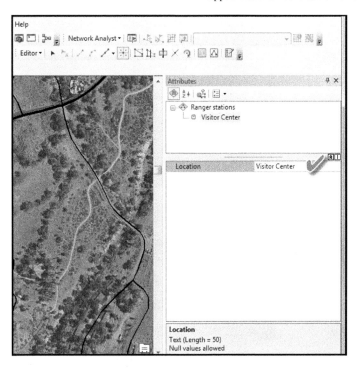

(*continued*)

ARCGIS LESSON 3-3 | (CONTINUED)

11. Take a screen capture to record your success so you can hand it in to your instructor.

Part II

Now that you have created and edited point features, take a stab at editing lines. In this case, you will be tracing a line representing a road. To do this, you'll use the same image as before, but there's more to it than before because the process has been started, so you have to use a feature called **snapping** that allows you to jump to the nearest edges or vertices or to other nearby elements. This allows you to make sure all the features you create are in the correct positions relative to other features.

1. If you have migrated away from your previous position, navigate back to the Editing folder and the Exercise1.mxd file. Remember to turn on editing and open the Create Features window as before. Now navigate to the bookmark called Digitizing roads.

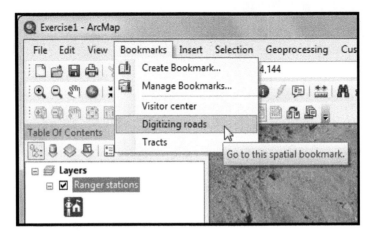

2. To use snapping functions, you need to turn on the Snapping toolbar from the Editing toolbar. *Note:* You can also add a toolbar from the Customize menu.

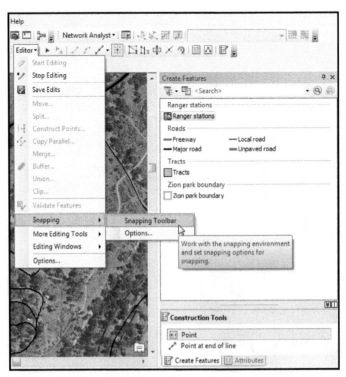

3. To make sure the Snapping toolbar tools are active, click the toolbar down arrow and make sure the tools are active (highlighted). If any are not, just click it to turn it on.

4. You also need to set some additional parameters. To do this click the snapping arrow and choose Options. A new window opens. Here are the options with which you should concern yourself right now. Ignore the others.

a. Tolerance: The distance within which a pointer or a feature is snapped to another location. Ensure that it says 10 pixels.

b. Snap tips: Text-based tips to help you while you work. Check all of them.

(*continued*)

ARCGIS LESSON 3-3 | (CONTINUED)

5. Click OK, and close the Snapping Options dialog box.

    You now need to enable the feature construction toolbar that gives you access to the common editing tools. As you become familiar with the process, you won't need to do this.

6. Click the Editor menu and select Options.

7. Check Show feature construction toolbar.

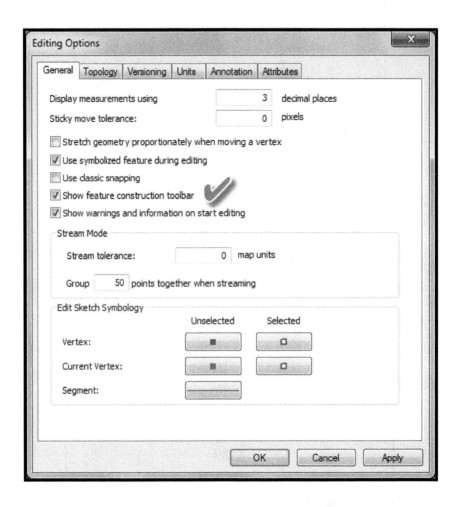

8. Click OK to close the Editing Options dialog box. You're now set to digitize the new road.

9. In the Create Features window, select Local Road. The construction tools appear. *Note:* It's a bit difficult to differentiate the symbols for Local roads and Major roads. If you want to save eyestrain, go to the Local Roads symbol on the Layers window on the left. Double click the symbol and change its color temporarily. Yellow gives a nice contrast for digitizing.

10. Notice that on the Local roads, near the very center of the map, there is a sharp right angle turn in the road, and a faint, undigitized road curving to the right. This is the portion of the road you will digitize. With the Line tool activated, locate the point where that right angle is located, but don't click just yet. Note that the cursor becomes a square. This is the Snap symbol. It should indicate the road endpoint.

11. Now click there to start and successively clicking one tiny section at a time, continue clicking along the road. *Note:* There is a very important concept here: The more curved the line, the more sample points you need to make. When you get toward the end, hover near the existing digitized road. This is where the snap point is that will join the line features.

(continued)

ARCGIS LESSON 3-3 | (CONTINUED)

12. There are several ways to snap your new digitized road to the existing road. Double clicking is the easiest, but you can also click the F2 key, or you can use the Feature Construction toolbar or right-click menu.

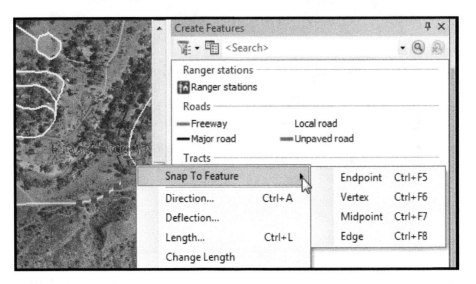

13. Whichever way you choose, you have now constructed your first line object in ArcGIS. Make a screen capture and submit it to your instructor.

Part III

In Part II, you finished digitizing a road and gained valuable experience with the editing interface. Now to create polygons, there's a bit of a story you need so you have some idea why you might want to create new polygons. While Zion National Park is mostly federal land, there are still some privately owned parcels, remnants of those owned before the land became a National Park.

1. Reopen the Exercise1.mxd and set up your edit session as before.

2. Turn off the World imagery layer in the table of contents.

3. Zoom to the Tracts Bookmark.

4. Click the Tracts template in the Create Features window. Note the default Polygon construction tool is the default.

Note also that the tracts share an edge between the Park and adjacent private tracts. This will help you construct your polygon. Also take a minute to double click the tracts symbol and change Ownership from <Null> to Private (you will have more opportunity to work with this information in the next lesson).

5. Make sure the Straight Segment construction method is selected on the Editor Toolbar. This ensures that the segments between each vertex will be straight lines.

6. Move to the right and snap to the upper intersection of the park boundary polygon and tract line feature. Click once.

7. Now move to the lower intersection and snap there. Click once.

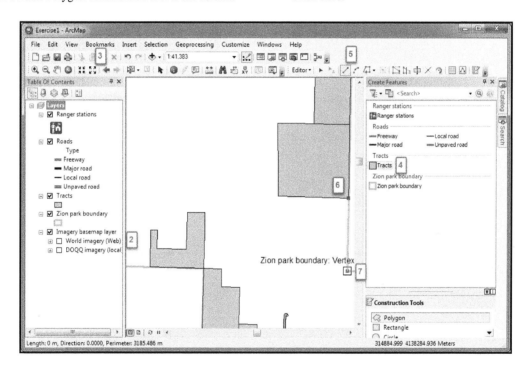

8. Click Midpoint on the Editor Toolbar pallet or the Feature Construction mini toolbar. This tool creates a vertex at the center between two selected points.

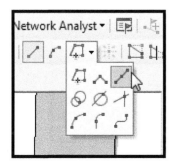

9. With that Midpoint tool selected, click the last location you selected.

10. Move to the left and click at the western end of the parcel polygon. Note the black line with the small square that appears in the middle of the line as you do this. The square shows you where the vertex will appear. When you click, the new vertex is added and a triangle-shaped polygon appears.

(*continued*)

ARCGIS LESSON 3-3 | (CONTINUED)

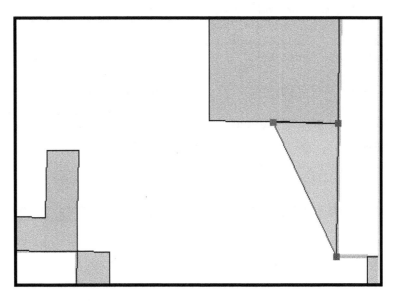

11. Now switch back to the Straight Segment method on the Editor Toolbar or Feature Construction mini toolbar. You are doing this because you are going to add one more point to create a more rectangular parcel but not by finding a midpoint. Instead, you're going to enter exact coordinates (another way to create feature locations).

12. Press the F6 key (the shortcut for Absolute X,Y), which allows you to type in the exact coordinates for the next vertex (by default the values are in map units, meters).

13. When the menu pops up, enter 314076.3 in the X box and 4138384.9 in the Y box as in the following figure. *Hint:* The escape key will cancel this entire operation, and the undo button will help you correct individual mistakes.

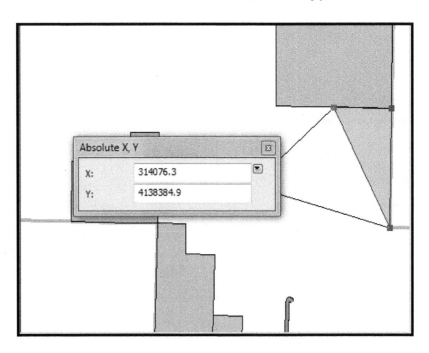

14. Finish the creation of the parcel by selecting the Finish Sketch button on the Feature Construction mini toolbar, or just press F2, resulting in the following.

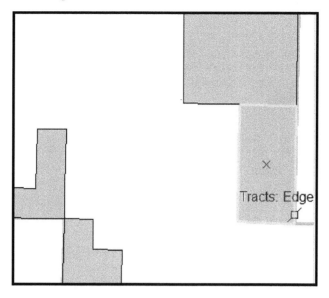

15. Recall that you were tasked with modifying the ownership of parcels like this in step 4. If you have done that, then this parcel should share that property. Click the Identify tool on the Tools toolbar.

16. Now click the new parcel you created. Notice the attribute value for Ownership is set to Private.

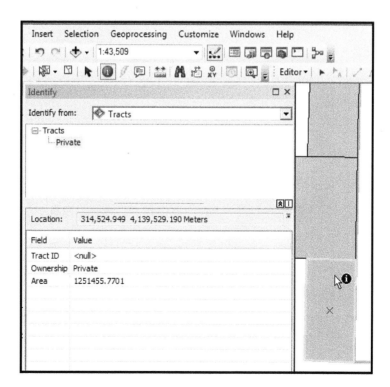

17. Take a screenshot of this to give to your instructor.

There are numerous other methods of creating polygons, such as creating true rectangles using the rectangle tool and adjoining polygons using the Auto-Complete tool. Try these out to see how they work.

Setting Feature Template Properties

Recall that all GIS data entry revolves around the template. Until now, you have pretty much been using the templates provided you. You are not limited by the use of either existing template layouts or existing templates. You can use the Template Properties dialog box to review existing settings and/or edit them to suit your needs. You can even rename a template, thus creating a new one. The new template contains any description, default construction tools, and attribute values for new features that users can create with your template. There is no review for this section. Move right into the ArcGIS lesson to practice making your own templates.

ARCGIS LESSON 3-4 | SETTING FEATURE TEMPLATE PROPERTIES

This short lesson gives you the opportunity to demonstrate ability to create your own template for input and editing.

1. Migrate to and open Exercise1.mxd in the Editing folder of ArcTutor (\ArcGIS\ArcTutor\Editing).

2. Go to the Tracts Bookmark.

3. Turn on editing in the Create Features window, and double click the Tracts feature template. This will open the Template Properties dialog box.

4. In the Description box, type Private Lands in Zion. The description appears when you rest your pointer over a template in the Create Features window.

5. Click in the Tags box and, right after Polygon, type a semicolon, a space, and then Zion, semicolon, and Landownership. Tags are used to search for templates so they are pretty important.

6. Ensure that the default tool is the polygon. If not, click the Default Tool arrow and select Polygon.

7. As in the previous exercise, Ownership should be set to Private.

8. Click OK. The template should look like the following.

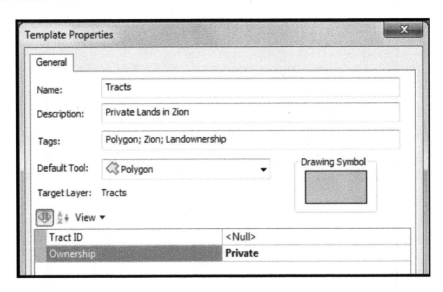

If you changed the name, you would effectively create a new template.

9. Return to the template and move your cursor over the Tracts name. The description you just entered now appears.

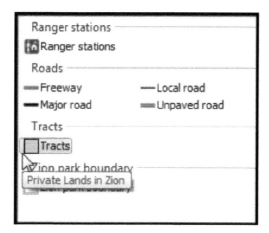

10. Take a screenshot, and give this to your instructor.

Creating and Editing Annotations

Annotation is a way to store text for placement on the map. For annotation, each piece of text stores its own locational information, text string, and display properties. The other method of placing text on maps is through the use of labels, which are based on feature attributes. How do you know which to use? If the position of a piece of text is important, you should store the text as annotations in a geodatabase. Annotation provides flexibility of appearance and placement of text because you have the ability to select individual pieces of text and edit it. The relevant point here is that you have the ability to convert labels to create new annotation features.

ARCGIS LESSON 3-5 | CREATING AND EDITING ANNOTATIONS AND DIMENSIONS

The Zion National Park layers you have been using in this chapter have dynamic labels, but some of the features aren't labeled because of space limitations. To get complete control over the position of text manually, you need to convert the labels to annotations.

Part I

1. Click the Open button on the Standard toolbar.

2. Navigate to Exercise3.mxd in the Editing directory (C:\ArcGIS\ArcTutor).

3. Select the map, and click Open.

Each of the feature layers has dynamic labels, and the Streams layer has label classes based on the layers' symbology. Label classes let you create different labels for different types of features for each layer. For example, intermittent streams can be given smaller labels than those for the perennial streams.

4. Click Customize, point to Toolbars, and then click Labeling.

(continued)

ARCGIS LESSON 3-5 | (CONTINUED)

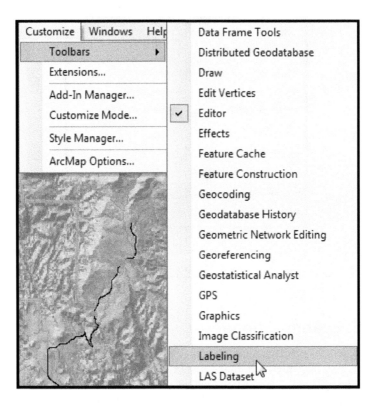

5. To get a feel for which labels don't fit, click the View Unplaced Labels button. Those labels that don't fit properly will be displayed in red. One rather tedious way to fix this would be to adjust the size, changing label weights or making the map larger. For this lesson, you will covert the labels to annotations and either place or delete labels that don't fit.

6. To hide the unplaced labels, click the View Unplaced Labels button again. You can toggle it on and off to see the differences.

While annotations are fixed in size and position (e.g., they get bigger when you zoom into the map), labels are drawn dynamically based on their properties and at a specified font size regardless of map scale. You can make labels act more like annotations, and you can set a specific reference scale for the map. As it turns out, when you convert labels to annotation, you also need to specify a scale or use the default map scale for the scale of the annotation. Zoom in and out on the map and observe what happens to the text.

7. Type 170000 in the Map Scale box on the Standard toolbar and press ENTER.

8. In the TOC, click the List By Drawing Order button (if not already set). Right-click Layers (data frame name), point to Reference Scale, and then click Set Reference Scale.

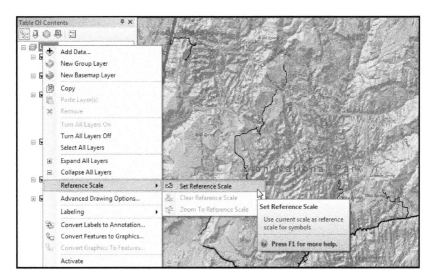

Now try zooming in and out and observe the behavior of the labels. At this point, you're ready to convert these to annotations. Annotations can be stored in either the map or geodatabase feature classes.

9. In the TOC, right-click Layers and then click Convert Labels to Annotation.

10. Uncheck all the Feature Linked boxes. Make sure that the Convert unplaced labels to unplaced annotations box is checked.

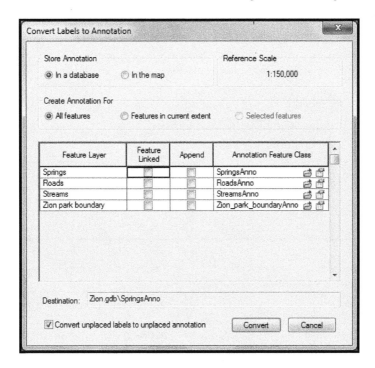

(continued)

ARCGIS LESSON 3-5 | (CONTINUED)

Notice that by unchecking these boxes, tiny folder icons appear. Before you unchecked the boxes, the layers' annotations, if created under those conditions, would be stored with the their associated feature class. Unchecking the boxes gives you the freedom to store the standard (nonlinked) feature classes in other geodatabases. By default, these feature classes are stored in the same dataset as their source feature class.

11. Click Convert. The labels are now converted to annotation. The annotation feature classes are now added to ArcMap.

Part II

Now that you have made the conversion, it's time to edit the currently unplaced annotation features and add them to the map.

1. Click the Editor menu on the Editor toolbar and click Start Editing.

2. Click the Editor menu again, point to Editing Windows, and then click Unplaced Annotation.

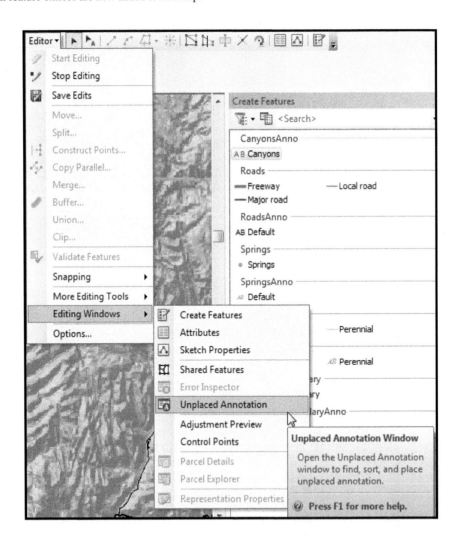

3. On the Unplaced Annotation window that appears, check the Draw box to display the unplaced annotation features on the map.

4. Click Search Now. A table of the unplaced annotation features of different classes appears. Additionally, on the map, you can see annotation features outlined in red.

5. Click the Edit Annotation tool on the Editor toolbar.

6. On the map, click and hold down the Z (zoom) key. Now using the cursor, draw a box around the cluster of unplaced annotation features on the east side of the park. You can also use the C key to pan, and if all else fails, navigate to the Zion Canyon bookmark.

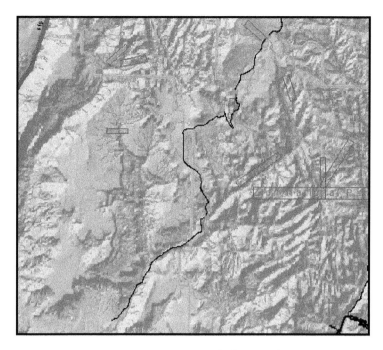

7. Click Search Now.

8. In the table, Birch Creek will be showing. Right-click on that column and click Place Annotation. Note that the Birch Creek feature is now placed and the box around it is blue instead of red. You have just placed your first annotation.

9. Take a screenshot and be prepared to give it to your instructor.

Part III

Now that you have placed your first annotation, recognize that it is straight and parallel to a segment of the stream feature while others are curved to follow the streams (a cartographic convention you'd like to maintain). The Follow Features option allows you to modify their placement and configuration.

1. With the Edit Annotation tool, right-click the Birch Creek annotation feature on the map, point to Follow, and then click the Follows Feature Options.

2. For the Make annotation, click Curved.

3. For Constrain Placement, click the Side cursor button to constrain the placement of the annotation.

4. In the Offset from feature box, type 100. This means that the annotation will be offset 100 meters from the stream in ground units.

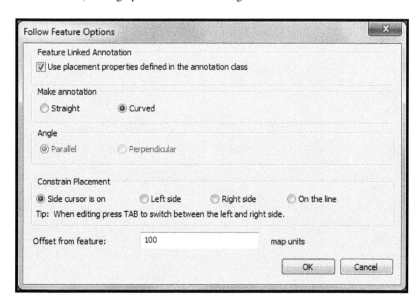

(*continued*)

ARCGIS LESSON 3-5 | (CONTINUED)

5. Click OK.

6. Now move the cursor over the stream feature just south of the Birch Creek annotation feature. Then right-click Follow This Feature. The feature flashes, and then the annotation curves along the stream segment. Drag the annotation back and forth until you are satisfied that it is placed where you like it. You can move it to either side of the stream as well.

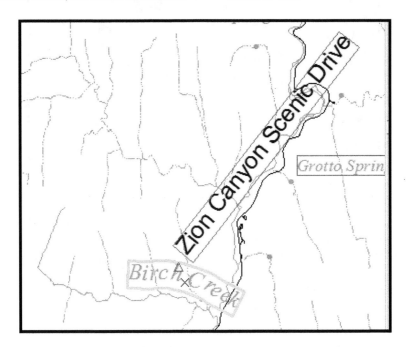

7. Make a screenshot of your success and be prepared to hand it in to your instructor.

There are other capabilities of the Edit Annotation tool.

8. On the Unplaced Annotation window, click Grotto Springs, then right-click it, and click Pan to Annotation.

9. Press the SPACEBAR, which is the keyboard shortcut to place a selected annotation feature. This selects the Grotto Springs annotation feature on the map.

10. Right-click the feature on the map and click Stack. Note that the text Grotto is now stacked on top of the text Springs.

11. Move the cursor over the middle of the Grotto Springs annotation feature.

12. Right-click Zion Canyon Scenic Drive and click Place Annotation.

13. Right-click the Zion Canyon Scenic Drive annotation feature on the map and click Stack again. The Grotto Springs annotation is now stacked.

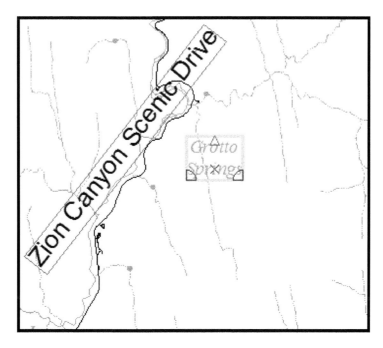

(*continued*)

ARCGIS LESSON 3-5 | (CONTINUED)

14. Move the pointer over the middle of the Grotto Springs annotation until the four-pointed Move Annotation pointer appears. Click and drag the annotation feature toward the southwest so it is between the spring features.

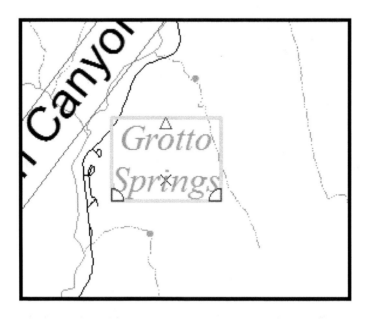

15. Click the bottom right-hand corner of the annotation and note the cursor change to a spiral arrow, meaning that you can now rotate the annotation.

16. From the right-hand corner, rotate the Grotto Springs so it is parallel to the Zion Canyon Scenic Drive annotation.

17. Take a screenshot and be prepared to turn it in to your instructor.

Part IV

One last thing you need to learn about Annotations is how to create you own.

1. Open the Create Features window by clicking Create Features on the Editor toolbar.

2. In the Create Features window, click the Canyons annotation feature template in the CanyonsAnno Layer.

When you activate an annotation template, the Annotation Construction window appears so you can enter the text and change the formatting of the feature you are going to create.

3. Click the Straight construction tool on the Create Features window.

4. Type Zion Canyon in the Annotation Construction window. As you type, the text on your pointer changes.

5. Click the map to the left of the road near Grotto Springs. The location you click is the center point of the new feature.

6. Rotate the annotation clockwise to align it with the road, stream, and canyon.

7. Click to place the annotation.

(continued)

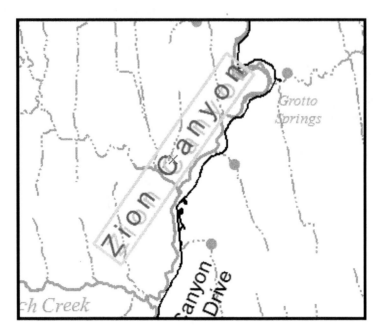

8. Press the E key until you have activated the Edit Annotation tool.

9. Place the pointer over the red triangle on the edge of the Zion Canyon annotation feature. The pointer changes to the two-pointed Resize Annotation pointer, allowing you to resize the feature.

10. Drag the resize handle toward the middle of the annotation feature. This forces the annotation to shrink as you drag it.

11. After resizing, drag the annotation if needed.

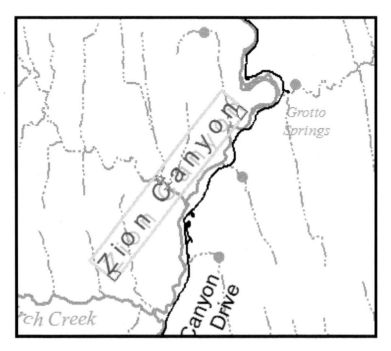

Now you need to create your own annotations that follow the edge of a line. This requires the use of the Follow Feature construction tool.

12. In the Create Features window, click the Default annotation feature template in the RoadsAnno layer.

13. Click the Follow Feature construction tool on the Create Features window.

14. On the Annotation Construction window, click Follow Feature Options to set options for how the annotation will be placed as it is dragged along. Be sure the annotation is set for curved and is constrained to be placed on the side the cursor is on and at 100 map units offset. Click OK.

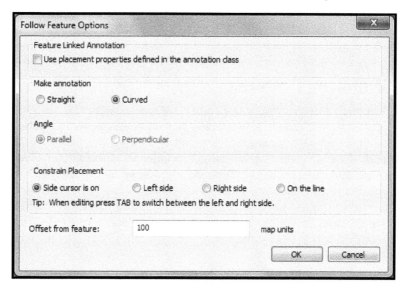

15. Click Find Text on the Annotation Construction window. This allows you to click a feature and populate the annotation string with an attribute from another feature.

16. Move the cursor over the road that branches off from the Intersection with Zion Canyon Scenic Drive and snap to and click the road. Highway 9 appears in the Text box on the Annotation Construction window and on the tool's pointer. If Zion National Park or Clear Creek appears, just press the N key to cycle through the text strings, or Find Text, move over the feature, and try again.

17. Click the road feature, which becomes highlighted, and drag the Highway 9 annotation along the line, pressing the L key if you need to flip the reading direction.

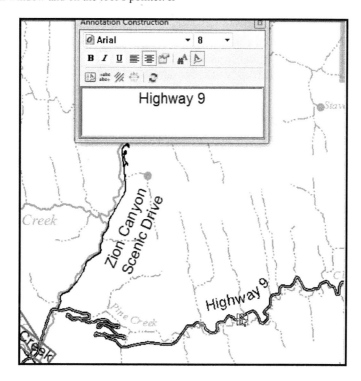

(*continued*)

---

18. Click to place the annotation.

19. Click the Editor menu on the Editor toolbar and click Stop Editing.

20. Make a screenshot of your work to give to your instructor.

21. Close ArcMap. There is no need to save your work.

Digitizing Features

It is really nice being able to analyze maps that someone else has already created, to be able to edit them, add features, and generally do what comes naturally. Fortunately, today's GIS market is filled with well-constructed, clean, relatively error-free databases. The sources for these vary as do the quality and the level of documentation (the **metadata**). For your own projects, the scale of available datasets may be too small, not cover your exact study area, not contain the layers you require; are in countries that don't have much available GIS data; or for some other reason, not conform to your needs. This requires that you create your own datasets or at least layers of existing datasets. Generally, the process for creating new layers is referred to as **digitizing**, and it usually comes in two forms: digitizing directly from hard-copy maps using a digitizing tablet or **heads-up digitizing**.

Heads-up digitizing might be considered more of an editing process than digitizing because you are creating new layers by adding features directly to on-screen layers. The most common approach to this is to have a digital version of an aerial photograph (e.g., an **orthophoto** or **orthophotoquad**). The features are clearly visible, and you use your mouse or other pointing device to trace and label the object right on screen. This method is quite nice because you can see your results and mistakes instantly. Editing becomes much easier as well because it can be done right away and interactively rather than after an entire map is digitized.

Every so often, though, you won't have digital data available to you for heads-up digitizing, and your layers must be created from scratch by digitizing the hard-copy document (map or image) using a digitizer tablet. Your instructor may have access to a digitizing tablet and some maps for you to digitize, but because not everyone has such a setup, for the time being, this chapter will discuss only the preparatory steps for digitizing. There are three basic steps that must take place: Preparing the map, registering the map, and enabling the digitizing mode.

Preparing the map actually requires first setting up the digitizing tablet and installing the software that drives it. For ArcMap, the digitizing tablet must have WinTab-compliant digitizer driver software. If possible, it is generally best if you can set up the digitizer first so the digitizer tool appears in the Editing Options dialog box. If it isn't, you need to register the digitizer .dll (digital link library) file using the ArcGIS EsriRegAsm.exe utility.

The second step in preparing the map is to register the paper map for digitizing. Registering a map provides the digitizer with a reference point for the corners of your map relative to its position on the digitizer. Registering also converts the coordinates from digitizer inches to real-world coordinates that include information about the datum and the projection used in the map you are digitizing. As your digitizer puck (a modified mouse) moves around the digitizer tablet, the coordinates change and can be registered using the buttons on the device. The buttons will be preprogrammed to recognize when you are digitizing points, lines, or polygons. Overall accuracy depends on the registration of the map coordinates, especially if you need to digitize the map in more than one session and the map must be removed from the tablet between sessions.

Finally, the third step, the enabling digitizing mode, provides you with an opportunity to operate in one of two modes. The **digitizing (absolute) mode** restricts your use to digitizing features. That means you can't choose buttons, menu commands, or tools from the ArcMap interface because the screen pointer is locked to the selected drawing area. In **mouse mode (relative**

mode), on the other hand, there is no correlation between the position of the screen pointer and the digitizing tablet. Fortunately, you can switch between absolute and relative mode using the Editing Options dialog box, thus giving you access to the interface choices as well as allowing you to digitize.

ACTIVITY 3-4 | **DIGITIZING FEATURES**

This exercise asks you some basic questions about what you need to know about digitizing.

1. Given the ready availability of so much GIS data, why does one need to know about digitizing?

2. List and briefly describe the three major steps needed to prepare for digitizing with a tablet.

3. What is the fundamental difference between absolute and relative digitizing modes?

ARCGIS LESSON 3-6 | DIGITIZING FEATURES

This ArcGIS lesson allows you to gain experience setting up a digitizing session. *Note:* If you don't have a digitizer attached to your computer running ArcGIS, feel free just to peruse this and remember it for future reference.

Part I: Setting Up the Digitizing Table and Installing the Driver Software

1. Close any ArcGIS applications that are currently running.

2. Start the DOS Command Prompt (found in your Windows accessories). *Note:* I'm using Parallels on an iMac, so don't expect to see the word iMac on your screen unless you are too.

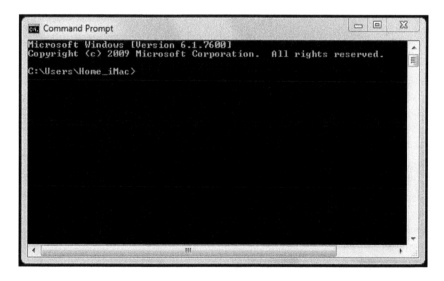

3. While in the Command Prompt window, type cd followed by a space then followed by the directory containing ESRIRegAsm.exe utility: C:\Program Files\Common Files\bin. This will change the command prompt's active directory to the folder containing the ESRIRegAsm.exe utility.

4. Press ENTER.

5. Then type ESRIRegAsm.exe, a space, an opening quotation mark, the full path to the ArcGIS installation location with the name of the DLL, and a closed quotation mark. The default path should be "C:\ProgramFiles(x86)\ArcGIS\Desktop10.3\bin\digitizer.dll".

(continued)

ARCGIS LESSON 3-6 | (CONTINUED)

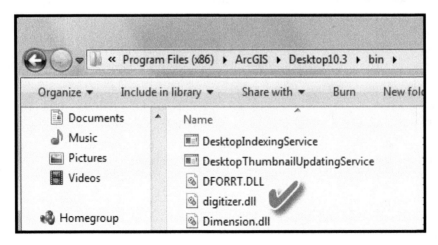

If you aren't using the default, that is, if you installed your software in another place, use that path instead.

6. Press ENTER.

7. If the registration was successful, close the Command Prompt window. The Editing Options dialog box will have the Digitizer tab when you next start ArcMAP.

Part II: Registering a Paper Map

1. Click the Editor menu and then click Options.

2. Click the Digitizer tab. Recall that your .dll file should be registered.

3. If this is the first time you are registering the map, do the following:

   a. With the digitizer puck, digitize the control points (essentially the X and Y locations of the map corners or perhaps the corners of the map boundary neat lines). The software will record the X and Y digitizer columns for each.

   b. Type the actual ground coordinates for each control point on the X Map and Y Map fields. The software will record the error in both map units and digitizer inches. To remove all the coordinates that you just entered because you made a mistake, click Clear on the Digitizer tab. If you want to remove a single record, just click the number in the point column for the coordinate you wish to remove and click Delete.

   c. If you are finished, click Save to save the ground coordinates for further use.

4. You can also register the map by loading coordinates from a text file.

   a. Click Load.

   b. Navigate to the file you want to use and click Open.

   c. Click the first record and digitize the first control point with the digitizer puck. The ground coordinates should appear in the X Map and Y Map fields.

   d. Digitize each of the remaining control points, noting the amount of error. *Note:* You do this to see if you are comfortable with the amount of error in your coordinates.

5. Click OK and your map is registered.

Part III: Enabling Digitizing Mode

1. Click the Editor Menu and click Options.

2. Click the General tab.

3. Type the **stream tolerance** (in map units). Stream tolerance is the minimum interval between **vertices**, usually measured in map units.

4. Type the number of vertices you want to group together. This setting is deleted by clicking the Undo button while digitizing in stream mode.

5. Click OK.

That's it. You are now ready to digitize your map. Your instructor may decide to include some digitizing for you to hand in.

Editing Topology

Topology is the study of geometric properties and spatial relations. In GIS, it forms a set of rules for how vector features that share geometry work together. You know, for example, that an urban feature is likely bounded by a suburban feature or a suburban feature is bounded by an agricultural feature. As represented in GIS, this means that the differently classified polygons share common edges. In a topological data model map, topology creates geometric relationships like this for any coincident objects. Not only does this explicit topology allow the GIS software to be much more efficient in searches, but it also assists in the editing of entities in the geodatabase.

ACTIVITY 3-5   **EDITING TOPOLOGY**

This exercise asks you some basic questions about topology.

1. Explain to a layman what topology in GIS is. Search the Internet, use the Esri site, or look up information in textbooks. Include diagrams if you can.

2. Do a WebQuest about the importance of topological versus nontopological GIS databases with particular reference to analysis operations and editing. Write a couple of paragraphs summarizing what you find.

ARCGIS LESSON 3-7 | EDITING TOPOLOGY

This is a quick example of how to create map topology and to use topology to fix line errors in a geodatabase.

Part I

1. Start ArcMap, and display Editor, Snapping, and Topology toolbars.

2. On the standard toolbar, click the Open button.

3. Navigate to MapTopology.mxd in the ArcTutor folder and open the file.

4. The map should look like this.

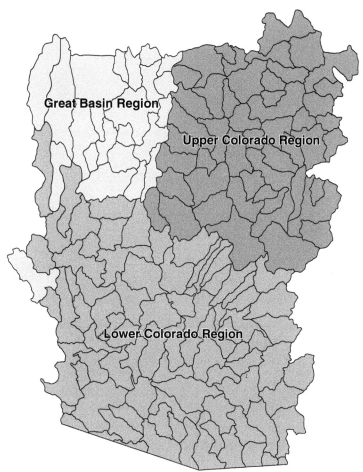

(continued)

ARCGIS LESSON 3-7 | (CONTINUED)

A bit about the map you just loaded. It contains two feature layers: hydrologic regions and watersheds (hydrologic units). The regions layer contains polygonal features representing three large hydrological units in the United States, each shaded with a different pattern. These features were derived from the smaller hydrological units, so the boundaries of the hydrologic region and those of the watersheds are already coincident. In the next part of this exercise, you are going to create map topology so you can edit the vertices that make up the shared edge at the intersection of multiple features.

5. Click the Editor menu on the Editor toolbar and click Start Editing. *Note:* To reduce the number of features that the map topology analyzes when building the topology cache, it is a good idea to zoom into the area of interest.

6. Click Bookmarks and click 3 Region Divide. The map now zooms to that bookmarked area with labels of the smaller watersheds.

7. Start an editing session, and select the Topology icon on the Topology toolbar. The Select Topology dialog box appears. There you can choose to select the feature to which you intend to add topology. The database doesn't have geodatabase topology, so it is grayed out. If the Options icon is not open, click its down arrow. The Cluster Tolerance shows and indicates how close together features need to be before they are considered coincident (adjacent).

8. Either select each layer's check boxes or use the Select All button on the right. This means both the Hydrologic regions and the Hydrologic units will participate in the topology operations.

9. Click OK. Now to start finding shared features.

10. Click the Topology Edit tool on the Topology toolbar.

11. Click on the edge that is shared by the East Fork Sevier, Utah (polygon #16030002) and Kanab, Arizona, Utah (polygon #15010003). The edge is selected and changes color. This edge is also shared by the larger regional

polygons. To check this, you will use the Shared Features window.

12. Click the Shared Features icon on the Topology toolbar. The names of both layers with checks next to their topology elements are shared and will be affected by any edits you make to the shared edge. Next, you will see which features share this edge.

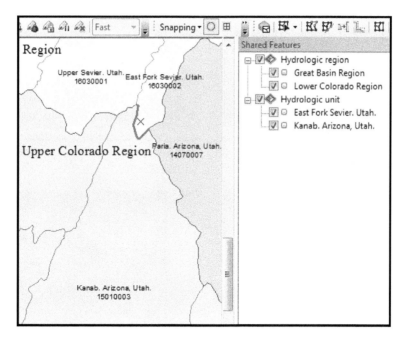

13. Click the Great Basin Region under Hydrologic region. It flashes on the map.

14. Click the East Fork Sevier, Utah under Hyrologic unit. It flashes on the map.

15. Close the Shared Features window.

Part II

Now that you've seen that the features you need to update share this edge, you'll update the boundary of the watersheds to better fit the terrain.

1. Turn on Hillshaded terrain layer from the TOC. This turns on a small portion of Hillshaded terrain from National Elevation Datased Shaded Relief Image Service published by the USGS. This will be used to update the watershed data.

2. Press and hold the Z key and drag a box around the selected edge. The National Hydrography data were compiled at a scale of 1:100,000. The hillside National Elevation Data were derived from the 1:24,000 scale digital elevation model data. You are going to use the higher-resolution hillside data to improve the watershed boundaries.

3. After you've zoomed into the area, double-click on the edge of the area. You will see the vertices (they appear in green) that define the shape of the edge you selected in step 10, Part I.

(continued)

ARCGIS LESSON 3-7 | (CONTINUED)

4. Move the cursor over the second vertex from the eastern end of the edge. When the pointer changes to a box with four arrows, click the vertex, drag it toward the northwest, and then drop it on the blue guideline.

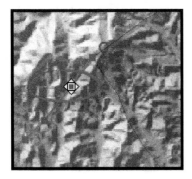

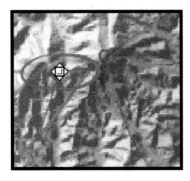

You could continue reshaping this edge vertex-by-vertex, but there is a faster way to update it.

5. Click once on the map off the edge to deselect it. Then click the edge again to reselect it.

Part III

In this part, you will use the edit sketch to reshape the shared edge. You'll need to use the Reshape Edge tool and snap to the watershed edges.

1. Ensure that edge snapping is enabled. If it isn't, click Edge Snapping on the Snapping toolbar.

2. Click the Reshape Edge tool on the Topology toolbar.

3. Move the cursor over the edge where the selected topology edge and the blue guideline begin to diverge.

4. Click the edge to start the edit sketch.

5. Continue editing vertices along the blue guideline. Make sure the last vertex snaps to the edge near the vertex you moved.

6. Right-click anywhere on the map and click Finish Sketch.

7. There are many other locations where you can reshape and modify edges. Do one more of these, take a screenshot, and give it to your instructor.

8. Click the Editor menu on the Editor toolbar and click Stop Editing.

9. Click Yes to save your edits.

10. Close ArcMap if you have finished working. You don't need to save the map document.

Part IV

Geodatabase topology is a set of rules defining how features in one or more feature classes share geometry. Geodatabase topology is created in the Catalog window or ArcCatalog and can be added to ArcMap as a layer just like any other data. After editing has been performed on the feature classes, you validate the geodatabase topology to see if the edits break any of the topology's rules. An ArcGIS for Desktop Standard or ArcGIS for Desktop Advanced license is required to create, edit, or validate geodatabase topology. Now you will get an opportunity to create a simple geodatabase topology rule to help you find digitizing errors in lot line data that have been imported from CAD and then use the topology and editing tools to fix these errors.

1. Start ArcMap and display the Editor and Topology toolbars.

2. Click the Open button on the Standard toolbar.

3. Navigate to the GeodatabaseTopology.mxd document and click Open. The map contains two layers. One layer is for parcel lot lines and the other shows your study area. You need to create a geodatabase topology so you can find and fix any spatial errors in the lot lines data.

4. If the Catalog window is not open, click the Catalog Window button on the Standard toolbar to display it. The Catalog window allows you to manage your datasets and is where you will add topology. You can dock the window to the ArcMap interface by clicking the pin in the upper right-hand corner.

5. If necessary, expand the Home - Editing\Topology folder, which displays the contents of the Topology folder installed with the tutorial data.

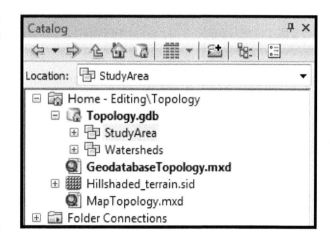

6. Expand the Tutorial geodatabase, if necessary, and click the StudyArea feature dataset. Your mission is to create a geodatabase topology to help find errors in the lot line data. To make life easy for you, the topology will be simple, involving one feature class and one topology rule.

7. Begin by right-clicking the StudyArea dataset, point to New, and click Topology. The New Topology menu shows up.

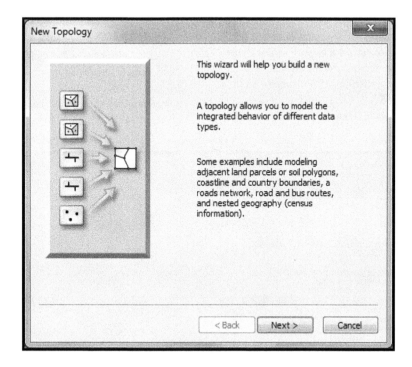

8. Click Next on the introduction. Here is where you can set cluster tolerance (minimum distance that separate parts of features can be from each other). Accept the default.

(continued)

ARCGIS LESSON 3-7 | (CONTINUED)

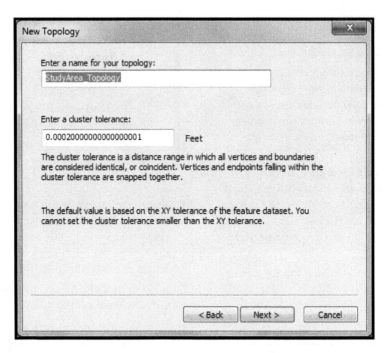

9. Click Next to select the feature classes you will include in the topology. Click LotLines and click Next. The next frame allows you to rank feature classes for editing if you have more than one feature class selection. They can range from 1 (highest) to 50 (lowest). These rules are used when decisions are needed to determine which feature is more important than others when editing.

10. Click Next and then Add Rule.

11. Click the Rule arrow and click Must Not Have Dangles. *Note:* Dangles are the endpoints of lines that aren't snapped to other feature classes. Sometimes they overshoot and sometimes they undershoot. These will impact the polygon count as well so they need to be corrected.

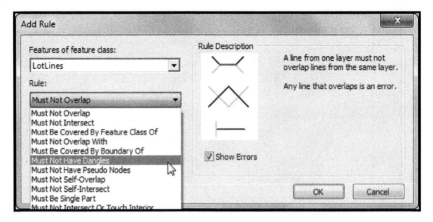

12. Now the rule is added to the list of topology rules. The summary report shows the new rule listed.

13. Review the summary report and click Finish. After a few seconds, the topology will be validated. The new topology now appears in the StudyArea dataset.

   Now you need to add the topology to the map so you can find the dangle errors. There may come a time when you

will need to create polygon lot features from these lines, so you need to have the correct number of polygons.

14. Go to the Catalog window and click StudyArea_Topology (you might need to expand StudyArea to find it). Click and drag it to the map.

15. When asked to add all the layers that participated in the topology, say no because the layers are already there in the map.

16. Close the Catalog window. On the map, the topology errors
    (point, line, and area) will show up. Because you were
    working only on dangles, only point errors show on the map.

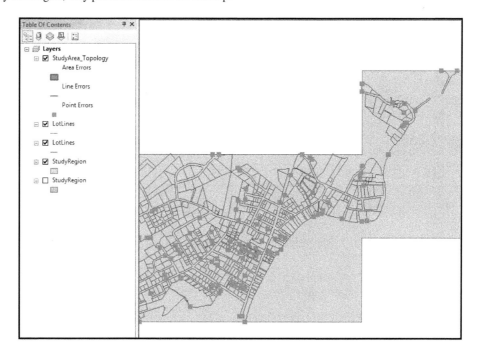

Part V

Now you need to identify the Lot line topology errors. Undershoots (lines not connected because they are too short) need to be
extended, and overshoots (lines not connected because they are too
long) need to be shortened.

1. Click Bookmarks and click Dangle errors so you zoom to the
   exercise study area where three dangles exist and need to be
   corrected.

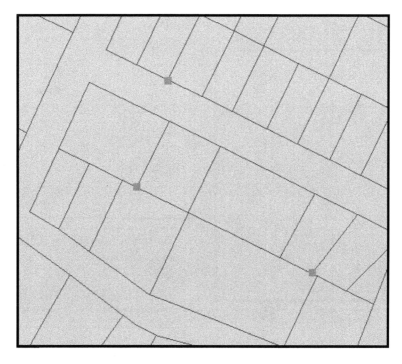

(*continued*)

ARCGIS LESSON 3-7 | (CONTINUED)

2. Click the Editor menu (Editor toolbar) and click Start Editing.

3. Re-adjust your screen if needed to see the same three errors on the inspector map as before.

4. Check the Errors and Visible Extent Only check boxes on the Error Inspector window.

5. Click Search Now on the Error Inspector window, and the table fills with the three data points to be corrected.

6. Click on the Feature 1 column and select Feature 144 (the northern-most error on your screen). Pan to center this and zoom in several times so you can determine the type of error. *Note:* This is an overshoot.

7. In the Error Inspector window, right-click Feature 144 and select Trim.

8. Type 3 in the Maximum Distance text box and press ENTER. The dangle is corrected and the error disappears from the Inspector window.

9. Click the Go Back to Previous Extent button (reverse arrow) on the Tools toolbar until you can see the two remaining errors. Pan if necessary.

10. Right-click on the lowest error showing in the Error Inspector window (Feature 180) and select Zoom To.

11. Zoom in a bit more until you notice the error.

12. You're going to use a different method of correcting this error. Select the Fix Error tool on the Topology menu, drag it to the error, and draw a box around it.

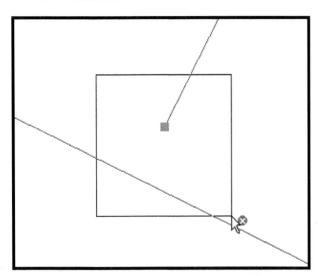

13. Right-click the map and select Extend.

14. Type 3 in the Maximum Distance text box and press ENTER. You have now corrected the undershoot error.

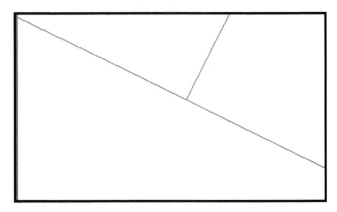

Now you're going to correct another error cause by digitizing a line twice.

15. Click the Go Back to Previous Extent button until you can see the remaining error.

16. Zoom to the remaining error.

17. Click Search Now in the Error Inspector window.

18. Click the numeric value in the Feature 1 column.

19. Zoom until you see the two parallel lines resulting from the double digitizing.

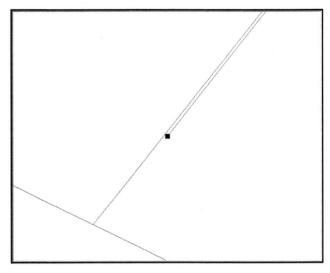

20. You should notice that one is dangling. To correct this, you will delete the line.

21. Right-click the numeric value in the Feature 1 column (Feature 182), and click Select Features (the dangle shows blue). Now click the DELETE key on your keyboard. The line is gone.

22. Click the Go Back to Previous Extent button and look at your handiwork.

23. For a better view, right-click Topology in the table of contents, click Properties, select the Symbology tab, and check Dirty Areas.

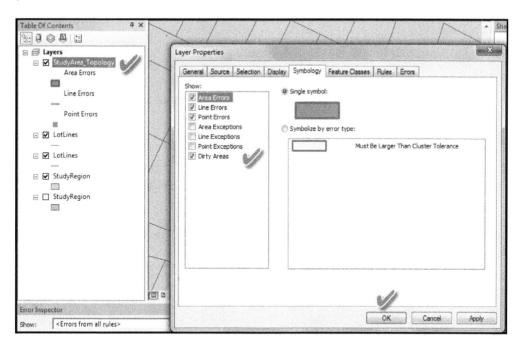

24. Click the Validate Topology In Specified Area button on the Topology toolbar. Drag a box around the northern dirty area. The dirty area is removed, indicating that your zone is error free. You can perform this for the southern zone as well if you wish.

25. To generate a report, right-click topology in the TOC, select Properties, select the Errors tab, and click the Generate Summary button. Get a screenshot of your error report to turn in to your instructor.

Rule	Errors	Exceptions
Must Be Larger Than Cluster Tolerance	0	0
Must Not Have Dangles		
LotLines	149	0
Total	149	0

26. Click OK.

There is a way to fix many errors at once to shorten the editing process, but be aware that some errors might not be real, so care must be taken.

27. Click the Extent button, then the Fix Topology Error tool, and drag a box around the entire study area, thus selecting all the errors. To remove all the undershoots, right-click the map, and click Extend. (Your maximum distance is already set from your previous editing steps.) If you repeat this process and then select Trim, you will remove the overshoots. The remaining edits are considerably fewer than what you had before.

28. Try this out, taking screenshots of the dirty areas to hand in to your instructor.

Principles and Practice of Spatial Adjustment

In the section Digitizing Features, you learned about how to create and clean typical line errors in a GIS database. The errors you worked with were primarily a result of data input issues, but there are many other data errors that need correcting, resulting from distortions in aerial imagery, mismatched edges on maps, and attribute errors. You already have a set of skills to help you, but this section will show you how to transform your data. These transformations involve locational movement, scale, changes, and rotation of graphic objects and text-based attribute manipulation. In the following activity, you will gain experience with using spatial transformations, rubbersheeting, edge matching, and transferring attributes among features.

ACTIVITY 3-6 | SPATIAL ADJUSTMENT

This activity gives you an opportunity to learn about data transformation.

1. Do a WebQuest to review the three basic types of graphic transformation: scaling, moving (shifting), and rotation. Provide a quick sketch of what each of these means.

2. WebQuest the term *rubber sheeting*. Define it. Explain the process (with a diagram if possible), and describe at least two situations in which it might occur.

3. WebQuest the term *Edgematching*. Define it. Explain the basic process (with a diagram if possible), and describe why this process might be necessary.

ARCGIS LESSON 3-8 | SPATIAL ADJUSTMENT

This lesson provides you with an opportunity to perform four types of spatial adjustment: transformation, rubber sheeting, edge matching, and transferring among attributes.

Part I

To begin, start ArcMap and display the Editor, Snapping, and Spatial Adjustment toolbars.

1. Click the Open button on the Standard toolbar.

2. Navigate to the Transform.mxd map document located in the ArcTutor folder.

3. Click and open the map.

4. Click the Editor menu on the Editor toolbar and click Start Editing. Set your snapping environment so that each link you add snaps to the vertices of features.

5. Ensure that vertex snapping is enabled. If not, click Vertex Snapping on the Snapping toolbar.

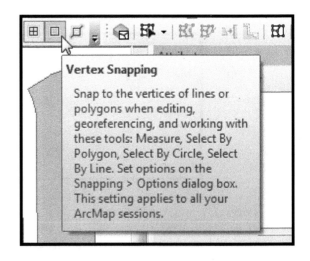

Vertex Snapping

Snap to the vertices of lines or polygons when editing, georeferencing, and working with these tools: Measure, Select By Polygon, Select By Circle, Select By Line. Set options on the Snapping > Options dialog box. This setting applies to all your ArcMap sessions.

To apply a transformation to your data:

6. Choose whether to adjust a selected set of features in a layer or all of them. Click the Spatial Adjustment menu on the Spatial Adjustment toolbar and click Set Adjust Data. Did you forget to open the Spatial Adjustment menu? If so, turn it on now.

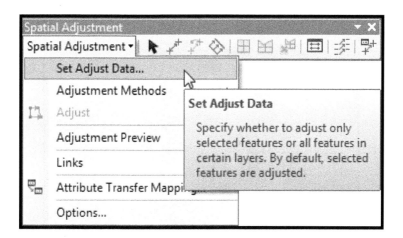

7. Click All features in these layers.

8. Uncheck the SimpleBuildings and SimpleParcels layers, keep the NewBuildings and NewParcels layers checked, and click OK.

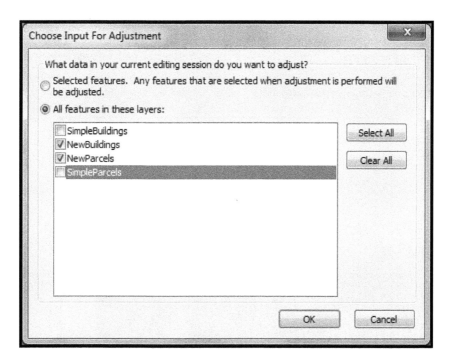

Now you will add displacement links (source and destination coordinates for adjustment). These can be created manually or loaded from a link file.

9. Click Bookmarks and then Transform.

10. Click the New Displacement Link tool on the Spatial Adjustment toolbar.

11. Snap to a from-point in the source layer and to-point in the target layer.

12. Continue to create additional links as shown below. For this ArcGIS Lesson, you will use a total of four displacement links when you are finished.

(*continued*)

ARCGIS LESSON 3-8 | (CONTINUED)

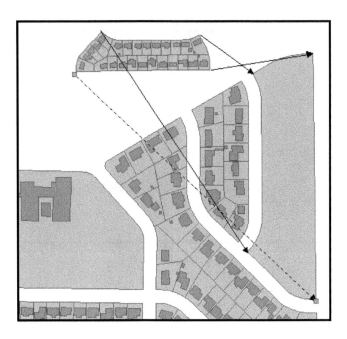

13. Click the Spatial Adjustment menu and click Adjustment Preview. A preview of the results shows, and if it doesn't line up, you can adjust the links as needed.

14. Click the View Link Table on the Spatial Adjustment toolbar. This provides information about the link coordinates, IDs, and RMS errors. Right-clicking a link record opens a shortcut menu where you can edit link coordinates. Both the Preview Window and Link table are there to assist you in your editing.

15. Click the Spatial Adjustment menu and click Adjust. The final product should appear like this.

16. Click the Editor menu on the Editor toolbar and Stop Editing.

17. Click Yes to save your Edits.

18. Close ArcMap.

Part II

Now you get a chance to practice rubbersheeting. Normally, this is used to align two or more layers that are meant to correspond. Imagine trying to line up four maps composed of land use data for four different time periods compiled from aerial photographs as an example.

1. Click the Open button on the Standard toolbar.

2. Navigate to and open Rubbersheeting.mxd.

3. Click the Editor menu on the Editor toolbar and click Start Editing.

   Set your snapping environment for each link you add snaps to the vertices or endpoints of features. *Hint:* Recall what you did in ArcGIS Lesson 3-8, Part I.

4. Ensure vertex snapping is enabled. If it is not, click Vertex Snapping on the Snapping toolbar.

5. Decide whether to adjust a selected set of features or all the features in a layer. Click the Spatial Adjustment menu on the Spatial Adjustment toolbar and click Set Adjust Data.

6. Click All features in these layers.

7. Make sure only the ImportStreets layer is selected. If needed, uncheck ExistingStreets and click OK.

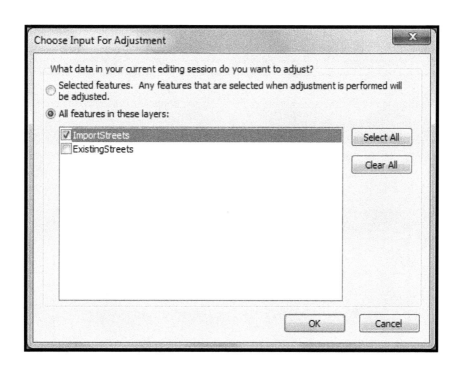

8. Click the Spatial Adjustment menu, point to Adjustment Methods, and click Rubbersheet to set the adjustment method.

9. Click the Spatial Adjustment menu and click Options.

10. Click the General tab.

11. Click Rubbersheet for the adjustment method so you can set additional options for rubbersheeting.

12. Click Options.

13. Click Natural Neighbor and then OK.

(*continued*)

ARCGIS LESSON 3-8 | (CONTINUED)

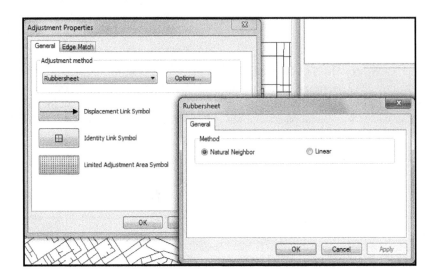

14. Click OK to close the Adjustment Properties dialog box.

   In the previous lesson, you learned about displacement links. For this exercise, you will create these links at several key intersections on the ExistingStreets and ImportStreets layers.

1. Click Bookmarks and click ImportStreets to get to the exercise view.

   Note that the ImportStreets (in red) and the ExistingStreets (in black) don't align. You are going to assume that the ExistingStreets layer is accurate, so you will be adjusting the ImportStreets by using Rubbersheeting.

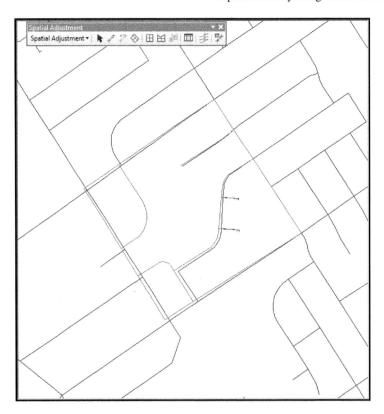

2. To get a closer look, zoom to the Intersections bookmark. This was prepared for you in advance. Feel free to zoom and pan until you are comfortable with the level of detail.

3. Click the New Displacement tool on the Spatial Adjustment toolbar.

4. Snap the link to the source location in the ImportStreets: Endpoint layer.

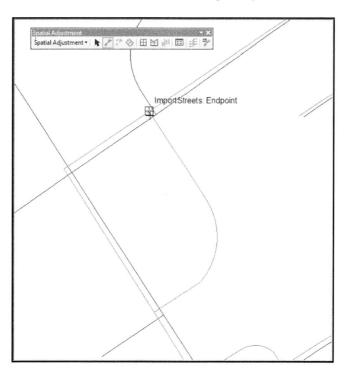

5. Snap the link to the destination on the ExistingStreets: Endpoint layer.

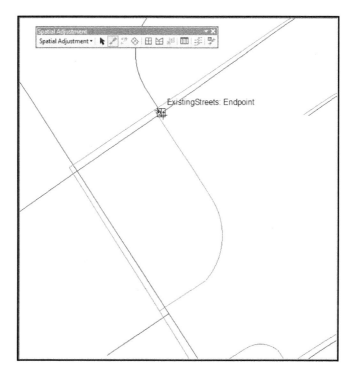

(continued)

ARCGIS LESSON 3-8 | (CONTINUED)

6. Continue this process in a clockwise fashion as in the following figure. You will create a total of six links.

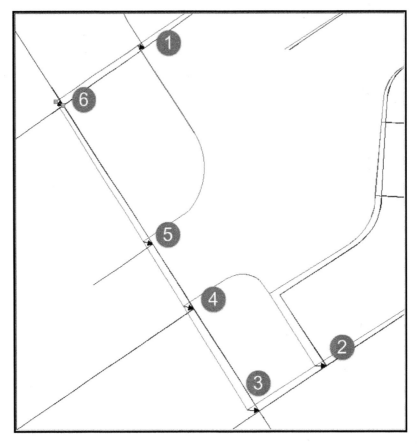

7. Zoom to the Curve Features bookmark to get a closer view. Zoom and pan as needed. *Note:* There are curved features that we haven't dealt with yet. You can add Multiple Displacement Links at critical points to make sure they keep their shape when you move them.

8. Click the Multiple Displacement Links Tool on the Spatial Adjustment toolbar to create multiple links in a single operation. This is a nice time-saving tool, especially for curved features.

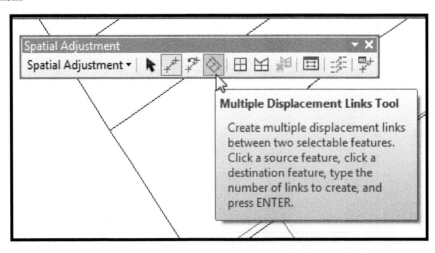

9. Click the curved road feature in the ImportStreets:
   Endpoint layer.

10. Click the curved road feature in the ExistingStreets:
    Endpoint layer.

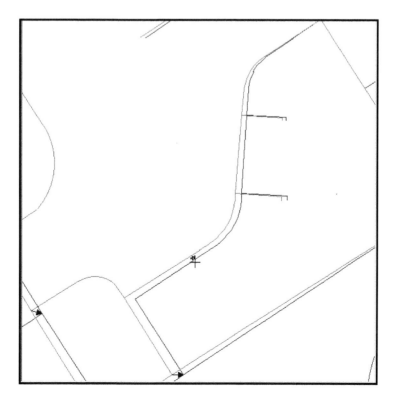

(continued)

ARCGIS LESSON 3-8 | (CONTINUED)

11. You'll be promoted to enter the number of links to create. Accept the default value of 10. Press ENTER. The 10 links appear.

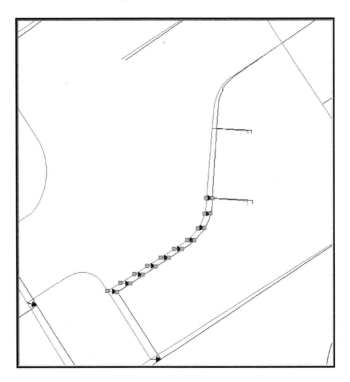

12. Create multiple links for the other curve feature to the north.

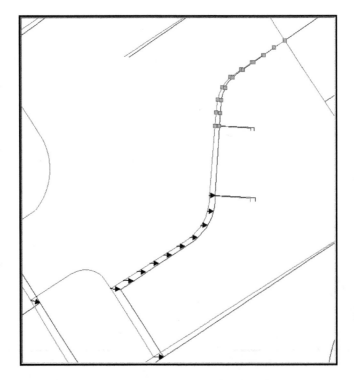

13. Click the New Displacement Links tool on the Spatial Adjustment toolbar.

14. Add the final displacement links.

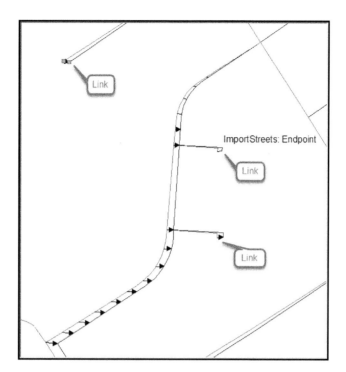

Now you need to add Identity Links that are used to anchor features at specific points to prevent their movement during an adjustment.

15. Click the New Identity Link tool on the Spatial Adjustment toolbar.

16. Zoom out and add five Identity Links at the intersections as shown below.

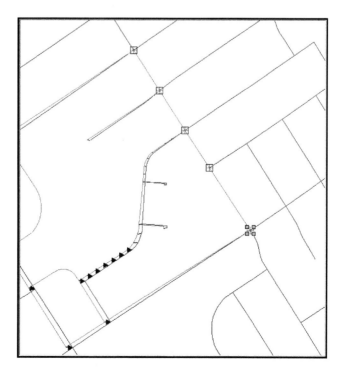

(continued)

ARCGIS LESSON 3-8 | (CONTINUED)

17. Click the Spatial Adjustment menu and then Preview
    to examine the adjustment. If your adjustments aren't
    satisfactory, you can modify the links.

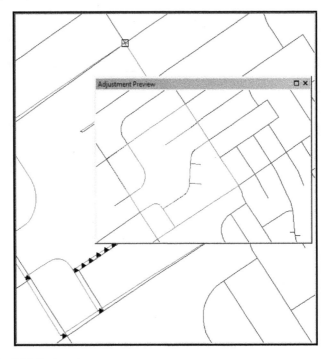

18. When satisfied, Click the Spatial Adjustment menu
    and click Adjust.

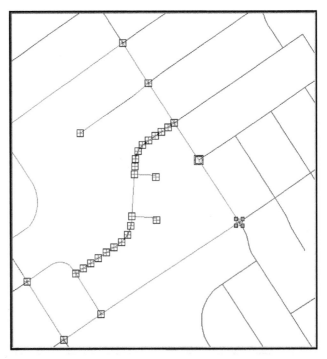

Notice that all the displacement links have turned into identity links. The next step is to delete these links because they are no longer needed.

19. Press the DELETE key on your keyboard.

20. Stop editing. Close the map.

21. Take a screenshot of your work to hand in to your instructor.

Part III

This section allows you to perform edgematching to adjust adjacent map layers. It is important to recognize that the map with the less accurate features is adjusted while the other layer is used as a truth set or target layer. Determining which is more accurate is often a judgment call based on your knowledge of how each map was created.

1. Click the Open button on the Standard toolbar.

2. Navigate to the EdgeMatch.mxd document in the ArcTutor folder and click to open.

3. Click the Editor menu on the Editor toolbar and click Start Editing.

4. Remember to ensure that end snapping is enabled. If not, Click End Snapping on the Snapping toolbar.

5. You will need to choose whether to adjust a selected set of features or all the features in a layer. Click the Spatial Adjustment menu on the Spatial Adjustment toolbar and click Set Adjust Data.

6. Click Selected features.

7. Click OK.

Now you need to determine which features will be adjusted. The next step is to choose an adjustment method. For now, you will use Edge Snap.

8. Click the Spatial Adjustment menu and select Adjustment methods, and then click Edge Snap.

9. Click the Spatial Adjustment menu and click Options.

10. Click the General tab.

11. Click Edge Snap for the adjustment method so you can set additional options for edge matching.

12. Click Options.

13. Click Line as the method and click OK.

The line method moves only the endpoint of the line being adjusted. The smooth method distributes the adjustment across the entire feature. The edge match adjustment method requires you to set properties that will define the source and large layers as well as to determine how the displacement links will be created when using the Edge Match tool.

14. Click the Edge Match tab.

15. Click the source Layer drop-down arrow and click StreamsNorth.

16. Click the Target Layer drop-down arrow and click Streams-South. The StreamsNorth layer will be adjusted to match the target layer, StreamsSouth.

17. Check One link for each destination point.

18. Check Prevent duplicate links and click OK.

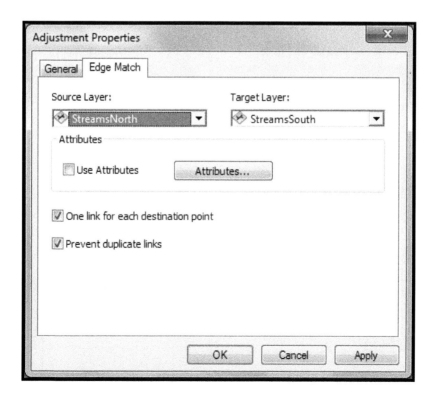

(*continued*)

ARCGIS LESSON 3-8 | (CONTINUED)

Now you are going to add edge match displacement links (remember these?).

19. Click Bookmarks and click West streams to set the view for editing.

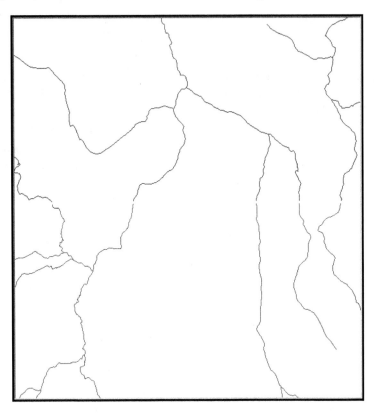

Remember that displacement links are used to define the source and destination coordinates for adjustment. In this section, you will use the Edge Match tool to create multiple links.

20. Click the Edge Match tool on the Spatial Adjustment toolbar.

21. Drag a box around the endpoints of the features. The Edge Match tool creates multiple displacement links based on the source and large features that fall inside the box.

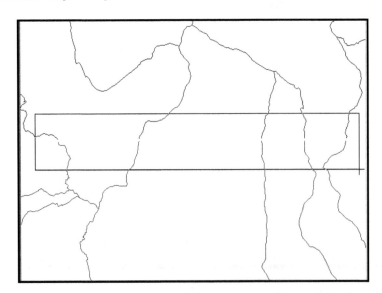

Displacement links now connect the source and the target features.

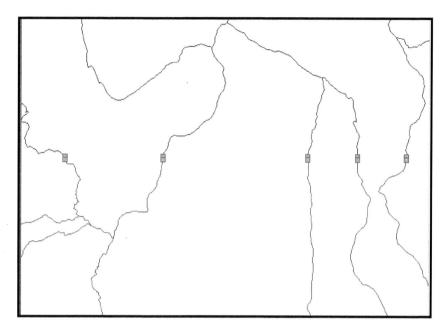

The edgematch displacement links form when the target and source features are within the preset snap distance. *Note:* The snap distance is based on pixels, so if you are zoomed in too closely, some of the links may not form. Zoom out and try again to correct this.

22. Click the Edit tool on the Editor toolbar.

23. Drag a box around the features that are to be edgematched. The links selected will be highlighted.

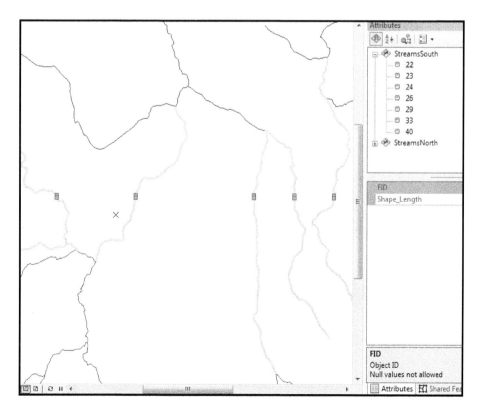

(*continued*)

ARCGIS LESSON 3-8 | (CONTINUED)

24. Hold the SHIFT key down until you finish adjusting the east streams. Click Bookmarks and select East streams.

25. Repeat the previous steps and create displacement links for the east streams.

26. Click the Spatial Adjustment menu and click Adjustment Preview.

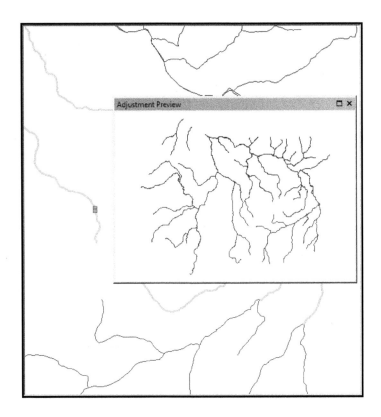

27. Click the Spatial Adjustment menu and click Adjust.

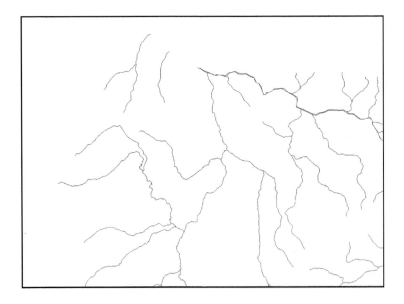

28. Take a screen capture for your instructor. Click the Editor menu on the Editor Toolbar and click Stop Editing.

29. Close ArcMap or move to the next part.

Part IV

So far, you've been focusing on the entities during your editing work. There are cases where attributes need to be moved from place to place, especially when working with source and target layers during edits. The tool for that is called the Attribute transfer, and this portion of the exercise will give you a chance to use it.

1. Start ArcMap and display the Editor, Snapping, and Spatial Adjustment toolbars.

2. Navigate to the AttributeTransfer.mxd map in the SpatialAdjustment directory of ArcTutor and click to open the map.

3. Start editing (in the Editor menu). Set snapping environment for both source and target layers to ensure that you select the correct features during attribute transfer.

4. Enable edge snapping.

5. Now you will set the source and target layers. Click the Spatial Adjustment menu and click Attribute Transfer Mapping.

6. Click the Source Layer dropdown arrow and select the Streets layer.

7. Click the Target Layer dropdown arrow and select the NewStreets layer.

   Now you will determine the attribute fields you want to transfer. You will select and match fields for source and target layers.

8. Click the NAME field in the Source Layer field list (left).

9. Click the NAME field on the Target Layer field list (right).

10. Click Add, and both appear in the Matched Fields list (bottom).

11. Repeat for the Type fields, and your form should look like this.

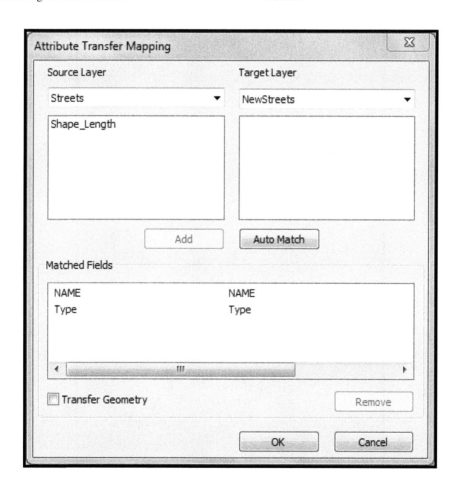

(continued)

ARCGIS LESSON 3-8 | (CONTINUED)

12. Click Bookmarks and go to NewStreets as the preset for your exercise.

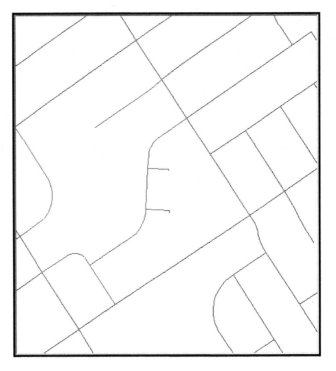

13. Verify the attributes of the source features by clicking the identify tool and clicking the source features.

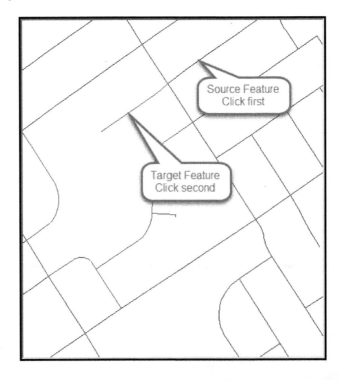

14. Notice the NAME (SAN PABLO) and Type 5 field attributes. The attributes are going to be transferred from the source to the target features.

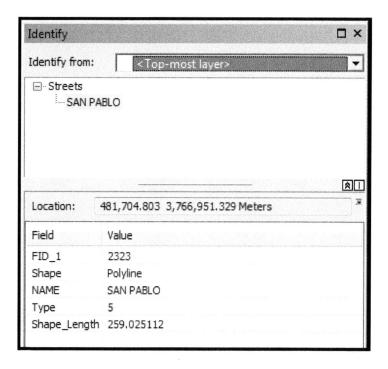

15. To see what's in the Target feature attributes, use the identify tool and click the target feature.

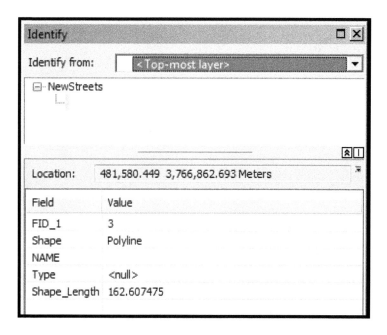

(continued)

ARCGIS LESSON 3-8 | (CONTINUED)

16. Notice that the NAME and Type fields are empty or null. You are going to transfer (copy) attribute data from the source to this target. Click the Attribute Transfer tool on the Spatial Adjustment toolbar.

17. Snap to an edge of the source feature.

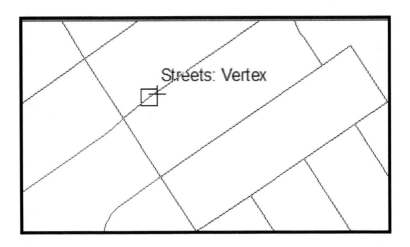

18. Drag the link toward the target feature.

19. Snap to an edge of the target feature and click.

20. To transfer the attributes, hold down the SHIFT key on your keyboard while selecting the target features.

21. To check your results, click the identity tool on the Tools toolbar.

22. Notice how the attributes are now transferred (copied) to the target feature.

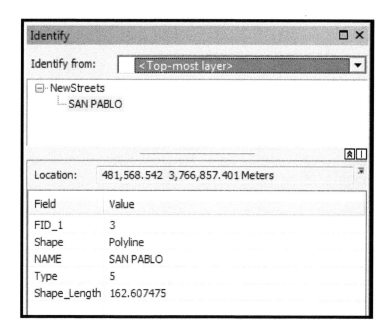

23. Take a screenshot of your final result to turn in to your instructor.

24. Stop editing and close ArcMap.

ADDITIONAL READING AND RESOURCES

Arctur, David, and Michael Zeiler. *Designing Geodatabases: Case Studies in GIS Data Modeling*. Redlands, CA: Esri Press, 2004.

Kerski, Joseph, and Jill Clar. *The GIS Guide to Public Domain Data*. Redlands, CA: Esri Press, 2012.

KEY TERMS

.dxf (Drawing Exchange Format): The file extension for AutoCAD by AutoDesk, Inc. for enabling data interoperability.

attribute: The descriptive information associated with map graphic objects.

COLLADA (Collaborative Design Activity): An interchange file format for graphic applications.

coverage: A georelational data model designed for storing vector data. It contains the spatial (location) data and the attribute (descriptive) data as a set of feature classes to represent geographic features. A major characteristic of the coverage model is that it contains topology that determines the relationships among the features.

digitizing: The process of converting traditional map or hard-copy imagery to digital form for use by the GIS software.

digitizing (absolute) mode: A mode of digitizing that restricts operations to digitizing features and does not permit access to other tools or menus.

entity: In a GIS, the graphic entity that defines the shapes and geographic locations of map objects.

geodatabase: An object-relational file structure used for ArcGIS that works across a wide range of database management systems' architectures and file systems. One major feature of the geodatabase is its ability to model behaviors of geographic objects.

heads-up digitizing: The digitization by tracing objects appearing as background on the screen with the GIS software.

KML (Keyhole Markup Language): An XML notation for defining geographic annotation and visualization designed for Internet-based earth browsers. Created by Keyhole, Inc., it was originally designed for Google Earth, which was originally named Keyhole Earth Viewer.

metadata: Data about data; a detailed breakdown of relevant information about the map layers in a GIS permitting the user to understand how best to use them.

mouse (relative) mode: A mode of digitizing in which there is no correlation between the screen pointer and the table area and which allows access to other digitizing functions.

orthophoto: An aerial photograph whose distortions are caused by differences in elevation from place to place have been corrected.

orthophotoquad: An orthophoto whose aerial extent corresponds to a quarter of a quadrangle map.

shapefile: An Esri nontopological vector storage format for storing the locations, shapes, and attributes of geographic features as a set of related files containing a single feature class.

snapping: The process of moving a feature to match or coincide exactly with another point or feature's coordinates when your pointer is within a specified distance (tolerance); commonly used to increase accuracy when using a variety of tools including editing, georeferencing, and measure tools.

stream tolerance: The minimum distance between vertices during a digitizing session. It determines how accurate your digitizing will be.

topology: The explicitly defined map geometry that tells the computer the specific geometric relationships among pint, line, area, and even surface objects. Topology allows for ease of search as well as entity editing in a GIS database.

vertices (singular vertex): The x,y coordinate pairs that define a line or polygon feature.

WFS (Web Feature Service): The interface standard that is an interface designed for operating on geographic features using platform-independent operators for Web-based GIS.

Basic Map Queries

CHAPTER 4

LEARNING OBJECTIVES

Here is the content you will learn in this chapter:

1. What is the basic organization of map layers in ArcGIS.

2. How to select layers.

3. How to identify the attributes of features on a map layer.

4. How to select individual or groups of features to examine.

5. Use of locational information to find features.

6. Use of feature attributes to select the right features.

7. Use of graphics to find features on the map (spatial query).

8. How to find features using addresses and place names.

9. The basics of building a query expression using the Structured Query Language (SQL) tools of your GIS.

10. How to use the SQL query tool to find features by attribute.

BEHAVIORAL INDICATORS

When you are finished with this chapter, you will be able to:

1. Describe and diagram the organization of map layers in ArcGIS, focusing on how this structure allows for retrieving and selecting appropriate map layers.

2. Demonstrate your ability to retrieve map layers using ArcGIS and provide explanations of what you have done.

3. Demonstrate your ability to identify the attributes of a variety of features on digital maps using ArcGIS and explain what you have done and how it relates to map reading.

4. Demonstrate your ability to use ArcGIS to select individual and groups of features and explain how this is accomplished and why it is useful.

5. Demonstrate the use of ArcGIS to isolate and select features based on their locational information and describe how this might be used in a GIS project.

6. Demonstrate the use of ArcGIS to select features based on the characteristics of their attributes and explain how this might be used in a GIS project.

7. Demonstrate the use of ArcGIS to select features by using graphics that determine where the features you wish to find are located. When you are finished, describe

how this approach is different from attribute queries, how it selects different features than an attribute query might, and how it is used in GIS analysis.

8. Demonstrate the ability to find features using addresses and place names using GIS, and describe how this might be used in GIS analysis.

9. Formulate a variety of structured query searches and describe what the query is designed to find.

10. Using ArcGIS, demonstrate your ability to implement several different examples of queries, provide examples of the output, and describe how this might be used in GIS analysis.

Chapter Overview

This chapter introduces you to the first, and most basic, of all the analytical capabilities of the GIS, that of being able to select which data will ultimately be used for more complex analysis. If you think about the basic purpose of GIS software as digital map analysis, the first thing you do when you encounter a collection of paper maps (essentially a hard-copy version of GIS) is to inspect them and decide which maps you want to look at and which specific details of the maps you need to view. Because GIS involves computers that are a bit more complicated than map drawers, you will first learn the basics of how the computerized map drawer stores and catalogs the maps so you can find them. Next you will learn a wide variety of ways of retrieving the information from these digital map layers. You will see that you can retrieve map information by individually pointing to objects, by searching for their names, by drawing graphic objects around the areas containing features you want to search, and by using attribute information as well as locational information. Among the more powerful techniques you will learn is how to develop and implement a search query using the Structured Query Language (SQL) built into the database management system and more specifically the **graphical user interface (GUI)** that the ArcGIS software uses to implement it. The ability to quickly, correctly, and efficiently search for and find the right data is an essential task for almost all the advanced analyses you will do in your career.

Essentials of Map Layers

Most modern GIS software has adopted an organizational structure in which the map is thought of as a **layer** of data. The data for that layer are usually rather specific in content, called a **theme** (a category of map data) because the types of maps included in the GIS are thematic maps. Typical thematic map layers might be soils, land use, hydrology, or topography. The layer is a representation of one of these sets of geographic data. You can think of a layer like a reference to the actual data that are stored in the attribute tables. For a layer to display or draw a map, the layer has to have direct access to the attribute data it is going to display. Layers also have to have a set of symbols and lines and marks to do the drawing. In the first case, most GIS software has a default set of such symbols. Frequently, these defaults are not very good, so after you select which layer you want to use, you will next have to decide which symbols or symbol sets you want to use for your selected layer (Figure 4-1).

In Figure 4-1, notice that the layer RoadL is present in the display window on the right because the layer has a check mark in the box. If you don't check that box, the layer won't appear in the display window. Additionally, if you select it and click the DELETE key or select Delete Layer from the pulldown menu, the layer will no longer be present in the list. Because RoadL was checked but nothing else has been changed, the display uses a default symbol set to draw the layer and its geographic contents for you. Clicking on the line symbol in the RoadL calls up a popup menu that allows you to select a series of different symbols so you can customize the map to suit yourself. After your new symbols have been applied, your map might look more like the one in Figure 4-2.

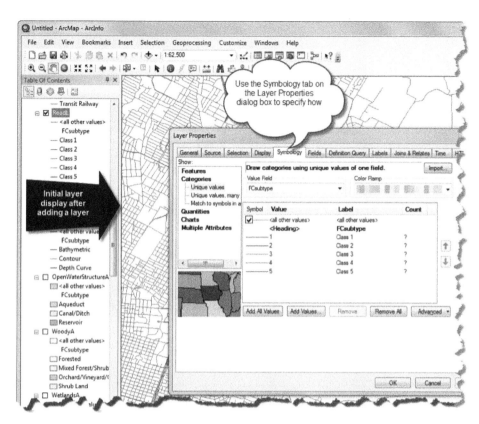

FIGURE 4-1 Initial display of layers in a GIS. The table of contents on the left indicates the presence of map layers. The legend also includes the RoadL layer with a check mark, indicating that it is active.

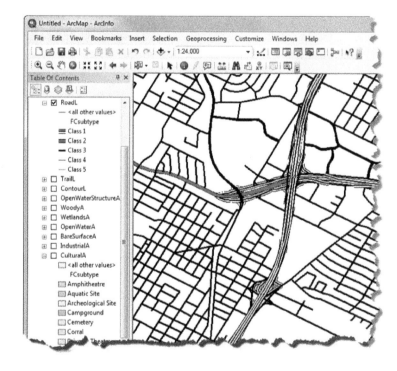

FIGURE 4-2 RoadL map layer with the customized symbol set applied.

In the following exercise, you will have a chance to explore an image of ArcGIS map layers, examine how the layers are displayed, select layers, view their default symbol sets, and modify the symbols to suit your needs.

Take a look at the following figure and answer the questions related to it.

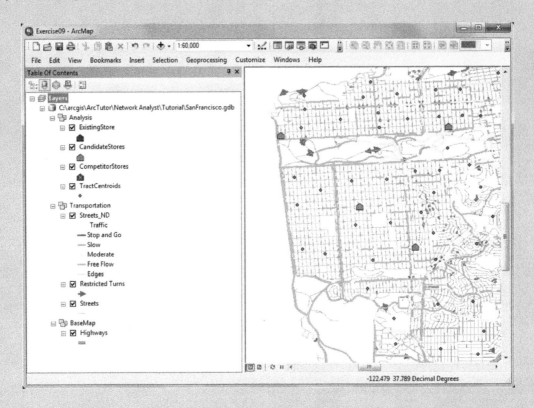

1. What is the source of the layer data based on the directory tree?

2. Which part of the figure is the map display area?

3. What are the names of the layers present in the database pictured?

   a. _____

   b. _____

   c. _____

   d. _____

   e. _____

   f. _____

   g. _____

   h. _____

4. Which layer(s) is/are active?

5. How do you know this?

6. The art galleries are symbolized by blue point symbols. Where would you click to pull up the symbol popup menu so you could change it to purple squares?

7. How would you change the color of the County polygon to pale blue rather than gray?

8. If you wanted to keep the layer called Agencies but not display it, how would you go about doing this?

9. How would you remove the Agencies layer entirely?

10. Would the action in question 9 eliminate the geographic data from your database?

11. What does the answer to question 10 tell you about what a layer is as opposed to a geographic data table?

12. If you click on a layer and nothing shows up in the display window, what might have happened?

ARCGIS LESSON 4-1 | EXAMINING MAP LAYERS

This exercise will acquaint you with the user interface of ArcGIS, particularly with regard to its ability to select or deselect a particular set of data layers from an existing database.

The student version of ArcGIS comes with a set of tutorial data. Using that data set, follow the steps and answer the questions. The questions are designed to determine whether you truly know what you are doing or are just following the directions.

Steps

1. Open ArcMap by clicking on the ArcMap shortcut or using the startup menu in Windows to migrate to the program. The following window will appear. As you can see, there are no layers selected and no maps displayed yet.

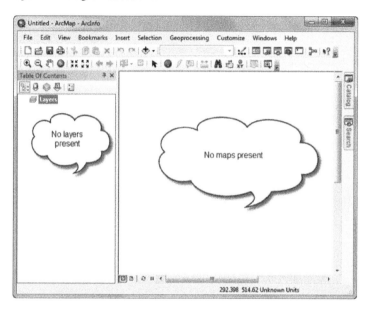

2. Click on the Add Data button ⊕, which is right below the Selection pulldown menu.

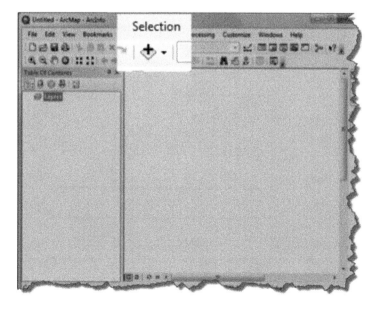

(continued)

ARCGIS LESSON 4-1 | (CONTINUED)

3. When you click the Add Data button, a popup menu appears with a list of possible data sets as well as a choice of navigation icons (see below). There are several ways to navigate through your folders. If you place your cursor over each of the icons, a popup text tells you what to do. For this lesson, you are going to connect to a folder because the tutorial data that you loaded from ArcGIS will be in a separate set of folders there.

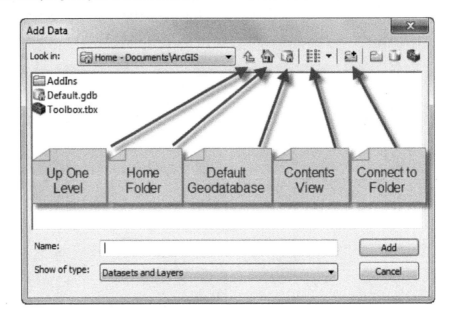

4. Using the Connect to Folder icon, navigate to the folder C:\arcgis\ArcTutor\Representations\Exercise_1.

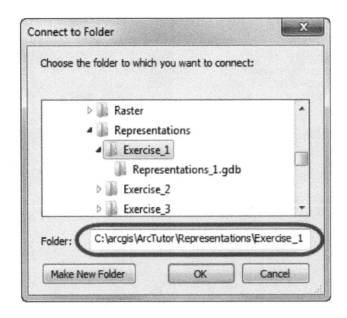

5. Click OK, and appropriate datasets and layers will appear.
   Select both of these.

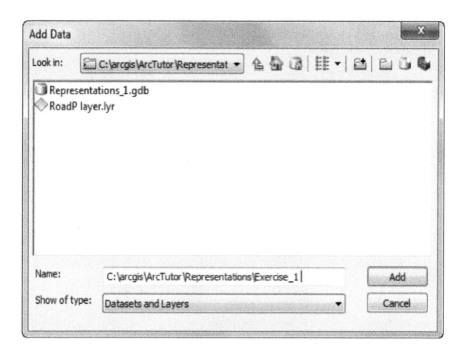

6. Select the .gdb (geodatabase) and .lyr (layer) files, and the
   following will appear:

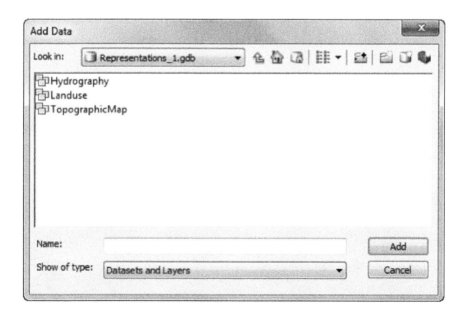

(*continued*)

ARCGIS LESSON 4-1 | (CONTINUED)

7. Highlight all three: Hydrography, Landuse, and Topographic Map. Select Add, and the following will appear:

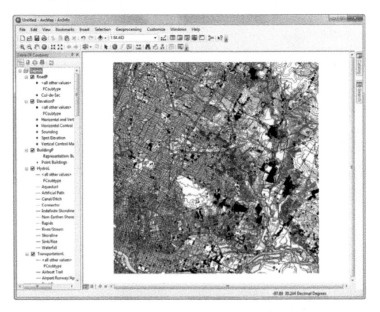

The table of contents on the left lists the map's layers. Each layer has a box next to it, and when it is checked, it means that the layer is active. Here, all of the boxes are checked, so all of the layers are active.

8. Double-click on the top layer, RoadP, which stands for Road Points. The popup Layer Properties menu appears with the default Symbology tab selected.

Selecting Map Layers

As you have seen, once layers are in your database, they can be displayed and their symbol sets modified to suit your purpose. Before you can do that, however, you need to be able to create a collection of layers from the entire database.

Using the Add Layers Button ⊕

In ArcGIS, the easiest and most obvious method of adding data is with the use of the Add Layers button, which looks like this: ⊕. Once you have clicked this button, the software will allow you to navigate through your computer file structure to locate the layers you want to add (Figure 4-3).

Additional Methods for Adding Layers

There are three additional methods of adding data layers. One takes advantage of the Catalog Window of the ArcCatalog module of ArcGIS, another uses the Search Window capability of ArcGIS, and the third uses ArcCatalog.

1. **Catalog Window of ArcCatalog.** Some users find it easier to search the layer catalog (catalog window) directly because it allows them to see the available datasets quickly and efficiently. Once the layer catalog is open, you can take advantage of the drag-and-drop capabilities of Microsoft Windows to select a layer you want, drag it to your open map or map window, and let it go to get Figure 4-4.

2. **Search Window in ArcGIS.** The Search window in ArcGIS provides another easy way to find just the right layers you want. By selecting Data just above the Search window and typing the name of the layer you want, you can again drag the new layer onto your map window (Figure 4-5). It is important to note that you *must* set search priorities before using the search. This tells ArcGIS which folders you want it to include in its search.

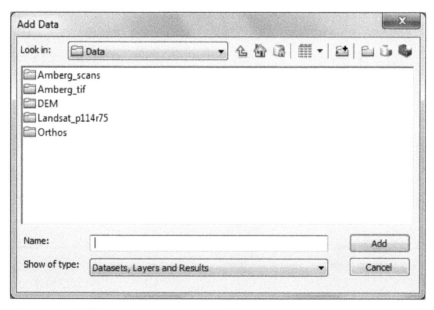

FIGURE 4-3   A typical GIS layer structure that the Add Layers button permits you to navigate. One thing you should know is that when you use this function the first time, it defaults to the last location from which you added data. You can change this option by customizing ArcMap (Customize > ArcMap Options > General).

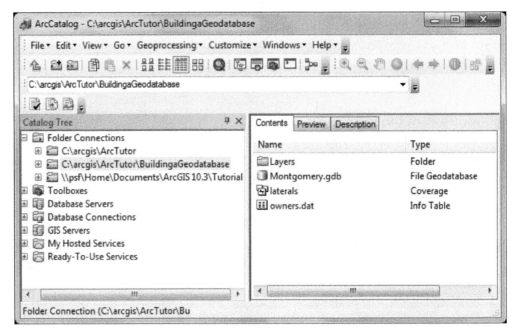

FIGURE 4-4 A portion of the catalog list showing available map layers that can be dragged and dropped into your open map database.

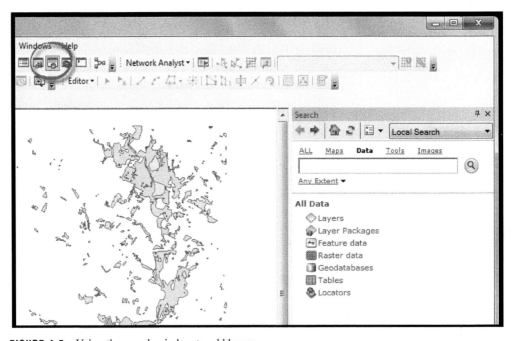

FIGURE 4-5 Using the search window to add layers.

ACTIVITY 4-2   SELECTING MAP LAYERS

The following simple exercises and questions review the different methods of adding map layers to your database.

1. In the following image, which method of map layer selection is shown?

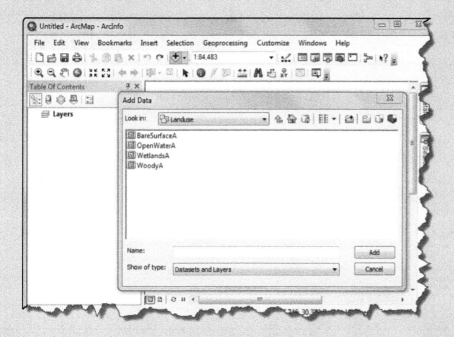

2. In the following image, which method of map layer selection is shown?

(*continued*)

ACTIVITY 4-2 | **(CONTINUED)**

3. In the following image, which method of map layer selection is shown?

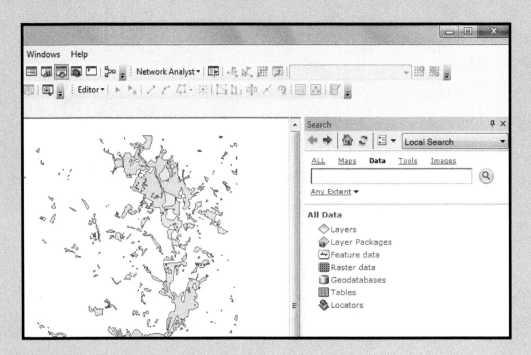

4. In the preceding image, if you got no return for your search but you know the data are on your computer, what step did you not complete prior to searching?

ARCGIS LESSON 4-2 | SELECTING MAP LAYERS

The purpose of this exercise is to give you an opportunity to experiment with at least three ways of selecting layers for your GIS activity. Follow the simple steps and answer the questions posed to you. The questions are designed to determine if you truly know what you are doing rather than just following the directions.

TIP

Review the procedures from ArcGIS Lesson 4-1 before going ahead.

Steps

1. Open up ArcGIS and navigate once again to Exercise 1 data in the ArcGIS tutor dataset.

2. As before, pull up all the map layers so that your map looks like the following one.

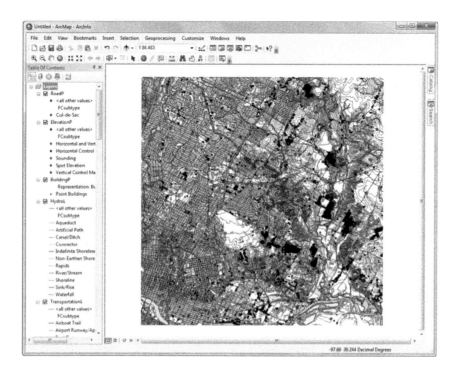

3. Navigate along the Table Of Contents and collapse the layers so that only the Layer title appears. This is done by clicking the tiny-sign inside the box next to the layer boxes with checkmarks. How many layers are present?

4. Delete all the layers that are not part of the Landuse category by highlighting each layer and clicking the delete key.

   a. How many layers are left?

   b. What are their names?

5. Check your work to see if you are left with only the Landuse layers. Clear your database and navigate to Exercise 1 again, but this time select only the Landuse layers and ignore Hydrography and Elevation. Why might you want to select only the Landuse layers?

6. Why might you want to have all the layers present?

7. Clear your database. Now employ each of the following:

   a. Add data layers to your workspace using the Add Data button in ArcMAP.

   b. Use ArcCatalog to add data layers to your workspace.

   c. Use the search window in ArcMAP to add data layers to your workspace.

8. Provide a screen capture of each of your methods of adding data to hand in. You should have X in all. Under each, provide a brief description of how you did it. Finally, describe below which method you prefer and why.

Identifying Features

You've now seen how you can include many layers in a current view of your overall database. After selecting one of the layers from your list and displaying the map, you see what the map reader would see: a more or less completed cartographic document. It would be nice, however, if you could delve a little more deeply to learn more about the specifics of the features in your map layers. To identify features in your data layers, you can either click the Identify icon ⓘ in the tools menu or right clicking and select Identify from the resulting menu. In the data frame, select the layer you want to examine by highlighting it. Then click the Identify icon and drag it over the map window that displays the data in your layer. When you do this, your cursor changes to become a little arrow with a black circle containing the letter *i*. Move this arrow cursor to a selection of features you want to inspect. Next, click and drag your cursor and draw a square. This will open into an Identify window with information about the thematic data for those features (Figure 4-6).

Alternatively, you can create a box that surrounds a group of features you want to examine by using the classic click and drag method. Or you can hold down the SHIFT key while selecting multiple features to view. Once you click on a feature with the Identify tool, the Identify window lists the features at that location. When you click on a feature, you will see its attributes in the bottom panel (Figure 4-7).

There are several ways to determine which layers you may select. The default for the Identify tool is the topmost layer of your map. You can change this by selecting Identify from the list located at the top of the Identify window to select:

- Topmost layer (the features highest up on the draw list)

- Visible layers (those layers that are currently visible on the map)

- Selectable layers (layers that you have set up as being selectable)

- All layers (whether they are displayed or not)

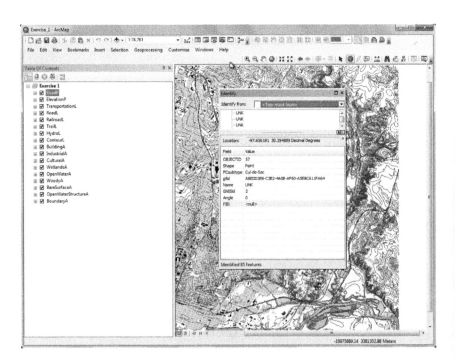

FIGURE 4-6 Menu resulting from using the identify function and creating a search window.

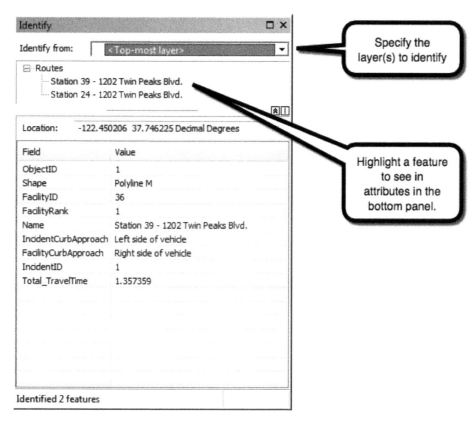

FIGURE 4-7   Attributes showing on the bottom of the panel after performing an identity search.

ACTIVITY 4-3   IDENTIFYING FEATURES

This short activity reviews what you know about using the Identify feature of ArcGIS. Refer to the following graphic.

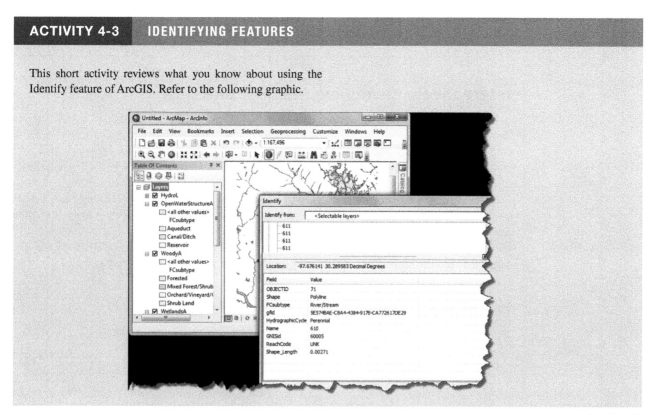

(*continued*)

ACTIVITY 4-3 **(CONTINUED)**

1. What types of layers is this identifying?

2. What are the other options under Identify from?

3. What types of features are shown in this figure, and what is their primary geometry?

4. Bonus question: What might the FC before subtype mean?

ARCGIS LESSON 4-3 | USING ARCGIS TO IDENTIFY FEATURES

This short exercise allows you to demonstrate that you can use the Identify tool effectively to find out what the thematic data are in your map layers.

Steps

1. Open ArcGIS and load the Hydrography, Landuse, and topographic data from the Exercise 1 dataset as in the previous exercises.

2. Use the tiny magnifying glass with the plus sign on the upper left of your screen to change your map scale to 1:63:360 (shown on the top center).

3. When you have finished, migrate to the very bottom right hand corner of the map. (*Hint:* You might have to move the scroll bars to do this.)

4. When you have completed that, select the Identify button and use the Selectable layers option.

5. Drag the Identify button to draw a square about 1 inch on a side and examine the resulting menu.

Explorations and Applications

1. Name the four primary data layers that appear in the top window.

2. Which layer has no data present?

3. What subtype is the one polygon selected from the BareSurfaceA layer?

4. Scroll to the top of the list and examine HydroL. What is the difference between the polylines numbered 611 and those numbered 624?

Interactively Identifying Map Features

Selecting layers from within the data frame is useful, but it is often handier to select features directly on the maps interactively because it is a bit more like working with paper maps. The first step in many analyses and for editing is to select the features that you want to work with, run statistics on, view, or use to compose a map document. It is more likely that you will use a subset of features rather than the entire set.

Generally, there are two methods of selecting features on a map: by clicking each feature one at a time with a mouse or by drawing a box using the drag and drop feature to surround the features on the map. The latter is called a **graphics-based search**. To select features interactively on the map, you can either use the Select Features tool 🔲 on the ArcGIS toolbar, or you can select data records from the data tables or the graph that are linked to the map graphics.

Preparing to Select a Set of Features in ArcGIS

In ArcGIS, the procedure to select a set of features first involves some preparation.

- **Select the layers.** The first step is to set the list of **Selectable layers** (layers that you can search interactively) by going to the table of contents and using the List by Selection view.

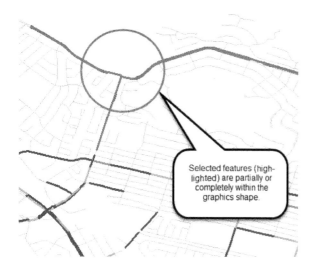

FIGURE 4-8 Using the select features that overlap in the Select Features tool.

- **Select the features based on three ways they can interact with your graphic.** Next you determine how the features are going to be selected based on how they interact with the graphic you will draw. The three options are:

  - o Select features that *touch or overlap* the graphic (Figure 4-8).

  - o Select features that are *completely contained within the shape* (Figure 4-9).

  - o Select the features that *completely contain the shape* (the opposite of above) (Figure 4-10).

  These options are selected using a pulldown menu accessed by Selection > Selection Options available from the main menu in ArcGIS (Figure 4-11).

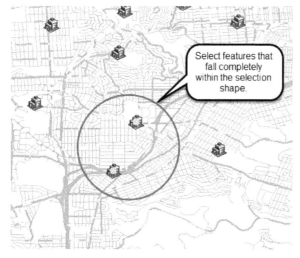

FIGURE 4-9 Using the select features that are completely contained in a graphic shape option of the Select Features tool.

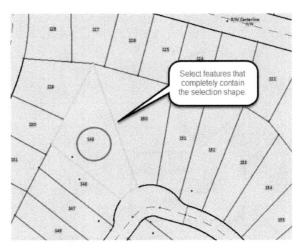

FIGURE 4-10 Using the select features that completely contain the graphic shape option of the Select Features tool.

- **Color options.** You can choose the color that selected items will show when selected by simply choosing the color from the pallet and the warnings related to the numbers of features selected from the list available. These will help you to know that you have the right features selected.

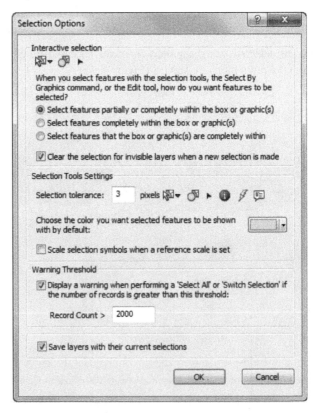

FIGURE 4-11 The Selection Options menu allows you to determine exactly how you want to search for attributes using the graphic search approach.

- **Clean or modified start options.** There are situations when you start clean and others when you might have already selected some features. You will be given the option of selecting a new set of features or modifying an existing set of selected features. Here are the feature selections from the Interactive Selection Method that you have:

  ○ Create New Selection

  ○ Add to Current Selection

  ○ Remove From Current Selection of features

  ○ Select From Current Selection from an existing set of features

  The menu for applying these options in ArcGIS is displayed in Figure 4-12.

Selecting a Set of Features in ArcGIS

Now that you have your options set, you are able to make the selection.

- **Click the Select Features arrow.** Your Selection Options will look like the one in Figure 4-13.

- **Select the desired features.** When you have made your selection, the features will show up as highlighted graphics based on the color you chose in your options selection (Figure 4-14).

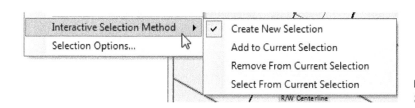

FIGURE 4-12   Options available from the Selection Options menu appear as a list.

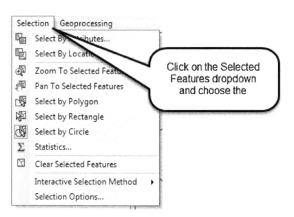

FIGURE 4-13   The Select Features arrow produces a pulldown list of options for different geometrical shapes you can use for your graphic searches.

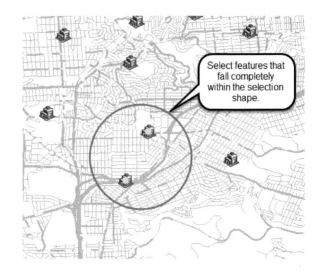

FIGURE 4-14 The highlighted graphics show which features have been selected.

ACTIVITY 4-4 | **INTERACTIVE FEATURE IDENTIFICATION**

This quick exercise provides you with an opportunity to describe how to use the Interactive Selection Method and what options are available. Examine the following figure and answer the questions related to it.

1. Where on the basic ArcMap menu would you find this menu?

2. In the HydroL features layer, which of the three options for feature selection do you think would be most appropriate? Explain your answer.

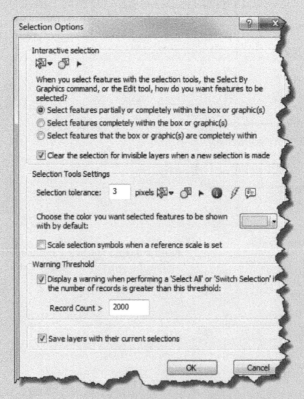

3. Where would you look to find the menu that allows you to determine the geometric shape that you will use to select by graphic?

4. What happens for a brief moment when you make a selection using the graphic selection method?

ARCGIS LESSON 4-4 | USING INTERACTIVE FEATURE IDENTIFICATION

This short exercise will allow you to practice selecting map items interactively.

Steps

1. As before, load the Landuse and Hydrography data from the Exercise 1 dataset. Make sure all the layers are loaded and visible.

2. From the Select By icon, try each of the five types: rectangle, polygon, lasso, circle, and line.

Explorations and Applications

1. Describe how each of the following works.

   Rectangle _____
   Polygon _____
   Lasso _____
   Circle _____
   Line _____

2. Speculate on why there are five types of geometry instead of just one. Why, for example, might you want to use one type of geometry over another? Give examples where you can.

Selecting by Location

There are times when you want to select features for a project because they are located in a particular place or near another set of features. Just as a quick example, you might want to select cancer mortality locations in proximity to a hazardous waste site. If there are more mortalities than might be expected, it would suggest a possible link between the cancer mortality rate and the presence of hazardous material. In ArcGIS, the Select By Location tool allows you to find features based on their location relative to features in another layer. The software provides you with a variety of methods to select point, line, or polygon features in one layer that are near or overlap with other features in the same or in another map layer. Figure 4-15 shows Select By Location when the target features touch the boundaries of the source layer features.

Figure 4-16 shows you how to determine the type of selection you want to make by clicking on the pulldown menu. This deals mostly with whether you are beginning a new search or are modifying an existing search just as you did in the Search By Graphic previously.

You will next have to select the target layers from which the features will be selected. Then you will select the geographic relationship you want to use to make your selection. Figure 4-17 illustrates a selection based on target features touching the boundaries of the source layer.

Finally, you will need to determine the source layer from which these selections can be made (Figure 4-18).

You will also have the option of using either the source features themselves in your search or, for some searches, you can use buffers (measured distances) to find your features.

Selected Spatial Queries

There are many options based on geographic properties from which to choose. These geographic properties describe how one set of features interacts with another set. Following is a list of the primary features available in modern GIS software and a general description of each.

- **Intersect.** Features either fully or partly overlap source features.

- **Are within a Distance of.** Uses a buffer (measured distance), returns features that intersect the buffers.

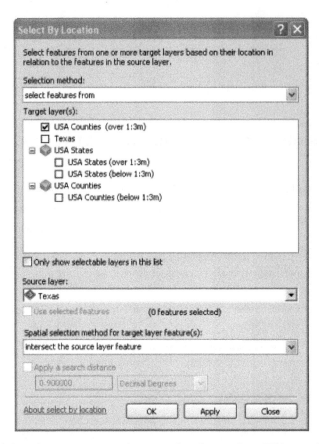

FIGURE 4-15 This is how the menu appears when you select features from USA counties where the Source layer is Texas and the selection method is Target layer(s) features touch the boundary of the Source layer feature.

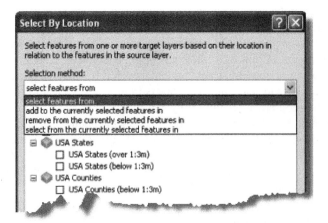

FIGURE 4-16 In the Select By Location pulldown menu, there are several options that appear as a list from which you can choose.

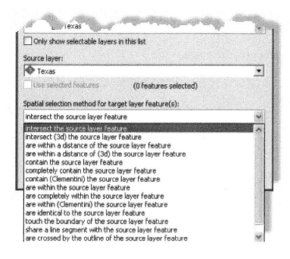

FIGURE 4-17 A list of the many options for geographic relationships from which you can search. In this case, the features are selected based on whether the target layer(s) features touch the boundary of the Source layer feature.

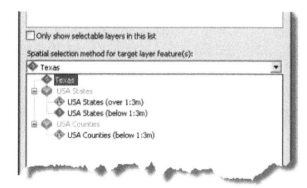

FIGURE 4-18 Source layers from which your selection can be made.

- **Are within.** Target feature falls within the source feature.

- **Are Completely within.** Target feature completely falls within the source feature and can't touch the source boundaries.

- **Contain.** Source feature must fall within the target feature and its boundary.

- **Completely Contain.** Source feature must fall within the target feature but cannot have common boundaries or overlap the boundaries.

- **Have Their Centroid in.** The centroid (midpoint) of the target feature must fall inside the geometry of the source feature.

- **Share a Line Segment with.** Source and target must share a common line segment.

- **Touch the Boundary of.** There are no interior intersections but there are places where their boundaries touch. This is called the **Clementini Touch Operator**.

- **Are Identical to.** All geometry (including edges) are equal.

- **Are Crossed by the Outline of.** Source and target share at least one edge, vertex, or endpoint in common but can't share a line segment.

- **Contain (Clementini).** Like **Contain**, but if the source feature is entirely on the boundary of the target feature and no interior portion of the source feature is inside the target feature, this operator doesn't select the feature while the traditional Contain operator does.

- **Are within (Clementini).** Like the **Are within** operator except that it returns nothing if the boundaries are entirely on each other but no part of the interior of the target feature is inside the source feature.

ACTIVITY 4-5 USING LOCATION FOR SEARCH

In the following brief exercise, you will have an opportunity to explain how Select By Location works.

Examine the following. Briefly describe what is going on. Include as much of the selection criteria as you can.

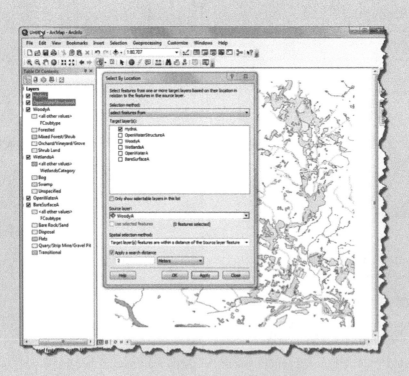

ARCGIS LESSON 4-5 | EMPLOYING LOCATIONAL SEARCH IN ARCGIS

This quick lesson will allow you to demonstrate that you are comfortable with the Select By Location option and can explain how it works.

Steps

1. As with the previous ArcGIS lesson, load the Landuse and Hydrography data for Exercise 1. (If a map is open from another session and other features are selected, click the icon to the left of the black arrow in order to deselect or clear all the features you previously selected from your map.)

2. Click on the Selection pulldown menu.

3. Observe that the first part of the menu asks you if you want to "select features from . . ." or to work with preselected features on the map. Since you have cleared previous selections, you can begin by selecting from the five target layers: HydroL, OpenWaterStructureA, WoodyA, WetlandsA, OpenWaterA, and BareSurfaceA.

4. Experiment by selecting one of these five as your target. Perhaps you might select HydroL so you can examine where streams are located relative to other features.

5. Select BareSurfaceA as your target because you are trying to see where the stream runs through bare surfaces.

6. Select "Target layer(s) features are within a distance of the Source layer feature."

7. Apply a search distance of 2 meters.

8. Click Apply and examine your results.

Explorations and Applications

1. Now, in a few sentences, explain the procedure you followed for this process.

2. The procedure is one thing, but the reasoning is quite another. Describe the goal of this procedure from a geographic perspective. In other words, what is the basic question the procedure is designed to help answer?

3. Bonus question: Propose a question similar to the one you just answered and employ similar procedures to achieve an answer.

Selecting by Attributes

Perhaps the most powerful tool for searching attribute data in a set of layers is to search for the descriptive attribute data stored in the data tables. ArcGIS has a user interface, a dialog box called the Query Builder. For searches involving attributes, use the Select By Attributes option from the selection menu in the general toolbar. The Query Builder window pops up, and you can search for any of the attributes you know exist by using the identify button in the previous section (Figure 4-19).

You should notice in Figure 4-19 that your first option is the layer you wish to examine. Once you have selected that layer, the software will know what attribute data are available for your query. The next step is to decide if this query is going to be a fresh query by selecting Create a new selection or if you are going to be working with a selection you have already started by selecting the Add to current selection option. Both of these are found in the second selection box labeled Method.

Whichever option you choose will result in a list of available fields (rows in your database tables) depending on the data available for that layer. Each field is a separate attribute, hence the reason that this is called an attribute search. From this list, you highlight the attribute you wish to evaluate ("MeterType" in Figure 4-19). It is a good idea to select the Get Unique Values button to ensure that all the values of the list will show. The only time you will *not* want to select that button is when you have a large number of values, such as with topographic data. When you select your value, a list of available attribute characteristics will display in the next window as well as in the bottom query window.

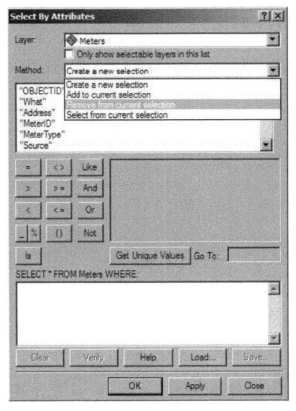

FIGURE 4-19 The Query Builder dialog box that appears when you select the Select By Attributes option on the main menu.

The first selection is only part of a larger set of selection criteria called an **expression**. In the example above, you selected "MeterType," but you haven't selected what type. You need to select an operator (something that operates on that selection). To determine which meter type you want, you need to select the = operator. When you do that, the = sign shows up in the query window. Next you select which characteristic of the attribute "MeterType" you wish. In this case, the selection was B. In this way, you can build more complex expressions. It is important to make sure that while you build your queries, each part shows correctly in the query box at the bottom. When the query expression is built, it is necessary to click Verify to make sure that the **syntax** of the query is correct so that it will select exactly what you want. When you are satisfied with the expression, click Apply to run the expression and let it do the search.

In the previous discussion, you had to use the = operator to select which properties of the attribute you wanted to select. You should also notice (see Figure 4-19) that the Query Builder Window contains a number of other operators. The operators provide the search relationships among the attribute characteristics. Figure 4-20 is what the operator search portion of the window looks like. Table 4-1 shows what each operator stands for and how it works.

The engine behind the Query Builder is a language called **Structured Query Language (SQL)** designed explicitly for building query expressions inside relational database management

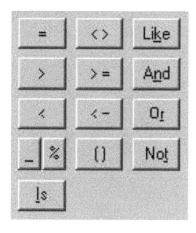

FIGURE 4-20 A portion of the Query Builder tool that contains the operators used in searches.

Table 4-1 List of Query Builder operators and their meanings

	ArcGIS Query Builder Operators
=	Equals
Like	Like equals but for character data and allows wildcards
>	Greater than (both mathematical and character data)
<	Less than (both mathematical and character data)
<>	Not equal to
>=	Greater than or equal to
And	Both expressions of a pair are true
Or	At least one of two or more expressions is true
Not	Excludes values. Normally used with "and"
Is	Used with databases that use the NULL keyword as a value for empty records.
( )	Parentheses used to control the order of expression evaluation
+, -, /, *	Mathematical operators
Wildcards (e.g., __ or ?) and % and *(all)	Used to refine an expression

systems such as those used in ArcGIS and many other GIS software packages. The Query Builder dialog box provides you with ample options for rather sophisticated queries, but you should be aware that as your skills grow, you can access still more power by using the **command line entry** of ArcGIS that is available in the ArcInfo GIS software.

ACTIVITY 4-6 | ATTRIBUTE SEARCHES

This brief exercise will give you an opportunity to demonstrate your mastery of searching by attributes. Examine the figure, and in the spaces provided after it, explain what it indicates. Describe the procedure in as much detail as you can.

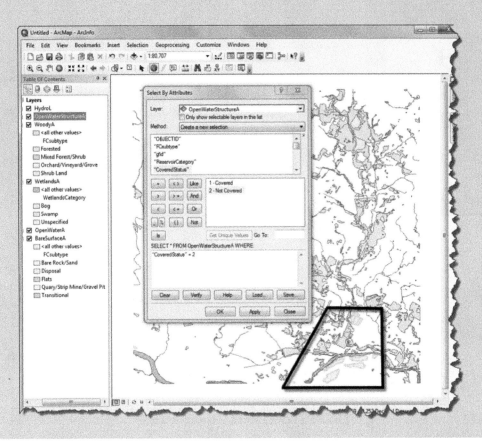

ARCGIS LESSON 4-6 | ATTRIBUTE SEARCHES IN ARCGIS

This short lesson gives you the opportunity to demonstrate your ability to use the Select By Attributes menu and understand what it means. Using the same datasets as before, employ the Select By Attributes to find all the covered sewage treatment ponds in the area.

Explorations and Applications

1. In the space below, write out the SQL query you used in the Select By Attributes menu to achieve your results.

_____

_____

2. Bonus question: Create a few additional scenarios, implement them, and turn them in to your instructor.

Using Addresses or Place Names to Search for Features (Find)

In map reading, it's very common to search for information by using either an address or a place name. If you've used Google Maps or Google Earth, you're familiar with this user interface. In a search strategy, you might search for a state, such as Rhode Island; a city, such as Kansas City; or even a known feature, such as Devil's Tower in Wyoming (Figure 4-21).

As with Google Earth, a full-blown GIS like ArcGIS also possesses this search feature. For ArcGIS, the interface is a tool called the Find tool. It is represented by an icon that looks like a pair of binoculars 🔍 and is found on the Tools toolbar. Unlike Google Earth, when you enter a search term in the search window, ArcGIS provides a list of features available in the layers stored in the GIS database that match the text string in your search (Figure 4-22).

Such a search results in a list of possible matches from which you can choose. As you navigate to each item in the list, it will be displayed as a map in the map window when highlighted. You can also double-click to engage the pan function, so it zooms to center the map on that location. Finally, you can right click to access a menu of applications that allow you to add markers, create a call-out label at the location, create bookmarks, or add it to a list of favorites and many more.

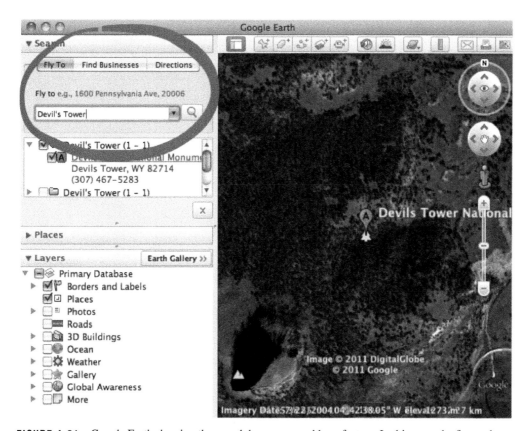

FIGURE 4-21 Google Earth showing the search by name or address feature. In this case, the figure shows a search for Devil's Tower.

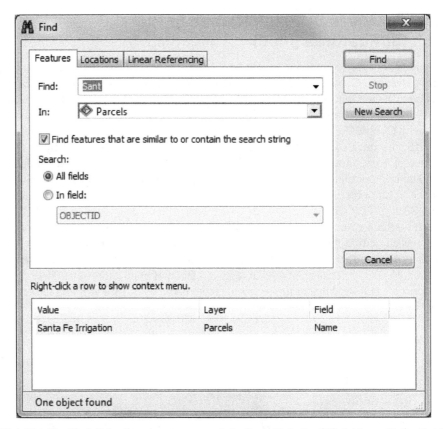

FIGURE 4-22 The Find dialog box showing a search for Santa Fe in the GIS database. Notice that the search is currently incomplete and has only the letters S-A-N-T, yet it pulls up features that match that letter string.

There are some options for the Find tool. You can, for example, find features that are similar or contain a search string. You can also select exactly which field within a layer you wish to search rather than just using the default settings.

The previous example used a text search based on name, but, like Google Earth and other simpler systems, you can also search for a specific address. In ArcGIS, this is done by opening the Find dialog box and selecting the Locations tab. Next you select a **locator** method from a pulldown menu in the dialog box. The default locator is ArcGIS Online (Figure 4-23), but other organizations such as the U.S. Geological Survey, U.S. Bureau of the Census, and many more have their own locators. These may focus on a specific region, and they are often designed to guarantee that the data are current and accurate. Once the locator is selected, you will then be able to enter the address of the feature you wish to locate.

The final option for searches includes **linear referencing**, a form of search using the Find tool in ArcGIS that allows searches on networks. Linear referencing allows you to find locations on networks based on their relative positions along measured linear features such as roads, rail lines, and paths. You might want to locate a car that had an accident at the 4-mile post marker north of a particular interchange. Or you might search for sample location number 43 along a belt transect used for identifying bird species.

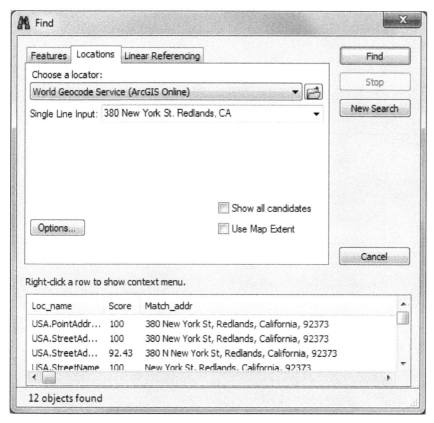

FIGURE 4-23   The Find dialog box showing the locator (in this case the default ArcGIS Online) plus an address search.

ACTIVITY 4-7   **ADDRESS OR PLACE NAME SEARCHES**

This exercise asks you some basic questions about searching for place names or addresses. When you have finished, the ArcGIS lessons will give you a chance to exercise your new knowledge.

1. What is the name of the ArcGIS tool that allows you to search a database by place names or addresses?

2. Describe what the icon looks like.

3. What is a locator?

4. What is the default locator in ArcGIS?

5. Explain what linear referencing is and how it is used.

ARCGIS LESSON 4-7 | PLACE NAME SEARCH IN ARCGIS

This is a quick example of how to do a place name search in Arc-GIS. Unlike Google Earth, which keeps all its data readily available in the cloud, ArcGIS requires that you first load the specific data from which you intend to perform your search. Follow these steps to see how this is done. When you have finished, answer the questions posed to you.

Steps

1. Start ArcGIS, making sure that all data from previous exercises are no longer active.

2. Go to the top left and click on File and then New.

(*continued*)

ARCGIS LESSON 4-7 | (CONTINUED)

3. A popup menu gives you a set of choices. In this case, select USA from the available template layers. The following image should appear.

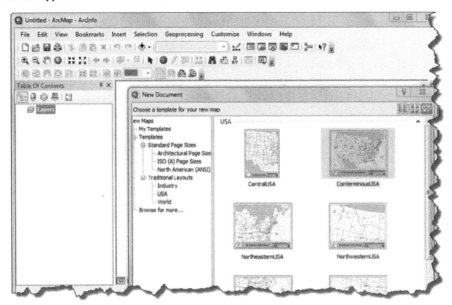

4. Once you have done that, select the North America template. The result should look like this.

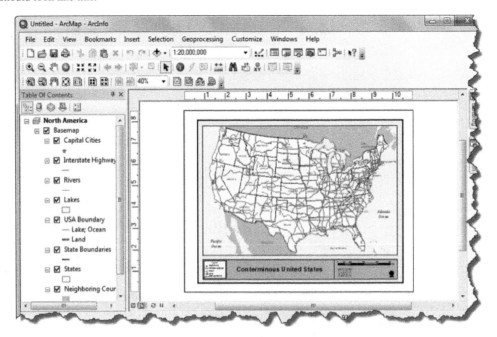

5. Notice that all the layers, including Capital Cities, are active. Click on the Find icon (the little binoculars). From the popup menu, select the Locations tab if it is not already selected.

6. Now select Choose a locator, which you might recall is Arc-GIS Online. Specifically select 10.0 North American Geocode Service (ArcGIS Online).

7. Type in the capital of South Dakota. (*Hint:* It's Pierre.)
   Your search should look like the following.

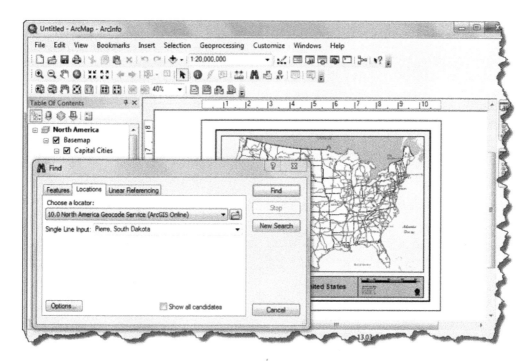

8. Click the Find button. Here's what you should see.

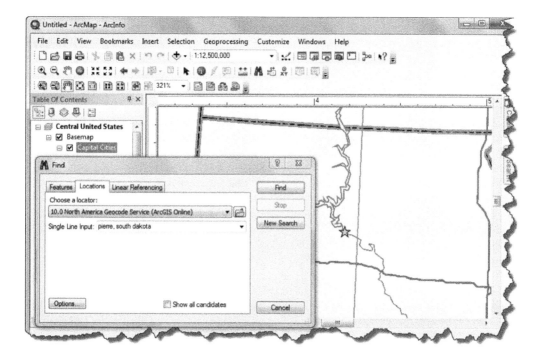

(*continued*)

ARCGIS LESSON 4-7 | (CONTINUED)

Explorations and Applications

1. Take a screenshot of your map and turn it in.

2. Perform three more searches, and turn them in with a short description of what you were searching for and what you got. If your search turned up unexpected results, speculate on what went wrong.

3. Explain what differences you noticed between Google Earth and ArcGIS both in terms of user interface and potential for analysis.

_____

_____

_____

_____

ADDITIONAL READING AND RESOURCES

Clementini, Eliseo, Paolino Di Felice, and Peter van Oosterom. "A Small Set of Formal Topological Relationships Suitable for End-User Interaction," in _Advances in Spatial Databases: Lecture Notes in Computer Science: Third International Symposium_, ed. David Abel and Beng Chin Ooi (1993), 277–95. A small set of formal topological relationships suitable for end user interaction.

RESOURCES

- A structured query language tutorial: http://www.sqltutorial.org/

- A bit dated tip sheet for selecting features in ArcGIS 9 but still quite relevant: http://ocw.tufts.edu/data/54/626829.pdf

- http://delab.csd.auth.gr/~alex/sdb/artSSD93.pdf

KEY TERMS

Clementini Touch Operator: A search operator that attempts to compare the level of containment of one set of graphic data with another but whose premise is that the interior and the edge of a polygon are fundamentally different entities.

command line entry: A method of entering commands that involves issuing text-based commands rather than with the use of a graphical user interface.

expression: A structured set of commands and operators within a Query Builder window of ArcGIS.

graphics-based search: A method of locating and selecting map features based on where the features occur on the map by drawing a graphic shape around the intended set of features.

GUI (graphical user interface): The set of menus, graphics, and dialog boxes used to support and enhance the interaction with software.

layer: A set of links to specific thematic map data in the GIS database.

linear referencing: A search strategy that finds a relative position along measured locations of a network.

locator: A set of directions for the computer to define the search for features using the Find command. Locators are predefined for specific organizations so they can control the extent of the database and ensure their accuracy and timeliness.

selectable layers: Layers that you designate as available for use by search dialog boxes.

Structured Query Language (SQL): A database computer language specifically designed to manage, control, and search relational database management systems. The ArcGIS Query Builder tool is based on SQL.

syntax: The rules and structure of the SQL designed to ensure that the queries are correct and to return the correct value.

themes: The general topics of thematic maps that map layers store, communicate, and are used for GIS analysis.

Common ArcGIS Tools

CHAPTER 5

LEARNING OBJECTIVES

Here is the content you will learn in this chapter:

1. Describe map overlay analysis and list some possible uses.

2. Identify the difference between Feature and Raster overlays and explain the pluses and minuses.

3. Describe proximity analysis and provide some possible uses.

4. List and describe some basic summary statistical analyses available in ArcGIS.

5. Describe how to access the Analysis Toolbox / Statistics Toolset.

6. Describe the utility of statistical analysis.

7. Describe what spatial statistical analyses are available in ArcGIS.

8. Illustrate how others have used statistical analysis in their work.

9. Describe basic table management procedures and capabilities of ArcGIS.

10. Describe how to perform some basic table management procedures.

BEHAVIORAL INDICATORS

When you are finished with this chapter, you will be able to:

1. Describe the general process of map overlay in GIS, explain at least three basic methods, and provide an example of how each might be used.

2. Describe the difference between Feature and Raster Overlays and briefly explain why one might be used over the other.

3. Describe what proximity analysis is and provide examples of how it might be used to solve geographic problems.

4. Perform a basic Buffer Analysis as an example of Feature-based Proximity Analysis.

5. List and describe the basic methods of nonspatial Summary Statistics available inside ArcGIS.

6. Provide a description of how to find and use the Analysis and Statistics toolsets.

7. Describe how statistical analysis might be used and what information you might derive from it.

8. Perform a WebQuest and provide at least three different examples of how others have used statistical analysis in GIS.

9. In your own words, describe some basic table management procedures and capabilities of ArcGIS and how they might prove useful.

10. Using your own words, describe how to perform some basic table management procedures.

Chapter Overview

This chapter provides you with some background information for some of the more common GIS tools available to you within ArcGIS. They are grouped here into Overlay, Proximity, Statistics, and Table Management, although one could also include other groups such as Surfaces. I've chosen to treat surfaces as a separate chapter so you can focus on tools normally reserved for 2-D analysis. The common term used is *geoprocessing* when considering the tools of GIS analysis. Although ArcGIS employs a graphical user interface, its structure and usage can be thought of as a form of language. The tools themselves are the action verbs associated with this language. So, for example when you think of the word *overlay,* consider that it is an action you are performing, not something you create (e.g., a mylar overlay is a noun while map overlay is a process).

Speaking of overlay as a process, not only is it a powerful tool, but also it is one of the reasons the first GIS software was produced. The concepts behind overlay are very geographical and involve the spatial associations of different variables across geographic space. For example, if you were to map the world's population, you would find that the distributions tend to be heavily clustered near sources of water. Such associations are considered when deciding what layers to overlay as the associations determine what patterns you wish to compare, contrast, combine, or mask. You may want, for example, to find all locations that are zoned for commercial use (based on a map of zoning) with those that are available to purchase (from a map of ownership) so you can find all available locations to purchase on which you can build a new business.

Proximity refers to the distance from one feature to another. You may want to find a house that is in proximity to a particular school or a playground or within easy walking distance to a grocery store. You may also want to locate farther from undesirable locations like industrial zones or sewage lagoons, which you can use proximity analysis to find as well.

Beyond measurements on a map, the document contains huge volumes of data about the features it contains. Some of these data, such as the number of available land parcels for farming, are explicitly defined. Others might need to be derived, for example, the average number of houses per acre in an urban area. Such quantities can be analyzed using the statistical capabilities of GIS. These capabilities might be based purely on the numbers of objects (aspatial) while others may require explicit knowledge of location (spatial). Both are easily calculated using ArcGIS statistics, such as mean, median, frequency, range, and standard deviation, as are all the statistical techniques normally available for Summary Statistics and the spatial equivalent of these techniques.

All of these data, of course, are contained in the attribute tables themselves from which a fair amount of manipulation is also possible. Within tables, you can rename categories, create new columns based on existing data or completely separate, or create features from columns containing coordinates. The manipulation of the tables is powerful by itself because of its ability to create features based on simple manipulation of other features within the tables. For example, you can create columns of population density by listing the population values in one column and the land area in another column. Being able to manipulate the tables is also useful for later spatial analysis as the new variables may be immediately used rather than having to be calculated during analysis.

Overlay Analysis

Generally, **overlay analysis** answers these questions: "What's on top of what else" and "Where is something on top of something else?" Specific questions might include the following:

- What types of vegetation occurs on a particular soil?

- Which houses are located in a 100-year flood zone?

- What roads run through areas with populations at risk like hospital zones or retirement villages?

- What archeological sites are located in a particular section of a national park?

There are many types of overlay analysis based on how the map layers are combined but they can, in the first case, be grouped into **feature overlay** (vector based) and **raster overlay** (raster based). It is generally easier to use raster overlay for overlay operations in which you are trying to find locations meeting selected criteria, although this is also possible to do this in vector. An advantage of raster overlay is that it has the map algebra capabilities (see Chapter 6) that allow mathematical combinations of grid cells while the vector overlay tools require primarily set-based operations.

Vector overlay is located in the Analysis toolbox in the Overlay toolset (Figure 5-1). Notice that there are seven different approaches to vector overlay appearing: Erase, Identity, Intersect, Spatial Join, Symmetrical Difference, Union, and Update. Erase does just as its name describes, erases part of an input polygon dataset based on the input features. Identity is a binary function (two layers only) that computes a geometric intersection of the input features and the identity features (Figure 5-2).

The parts of the input features that overlap the identity features obtain the attributes of those identity features (Figure 5-3). In the Intersect method, those features that are common to the two layers are retained, while in the opposite approach—union intersection—all input features are retained (Figure 5-4).

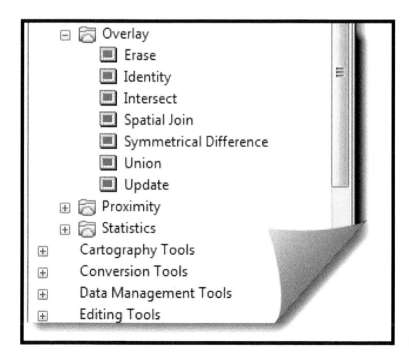

FIGURE 5-1   Overlay toolset user interface.

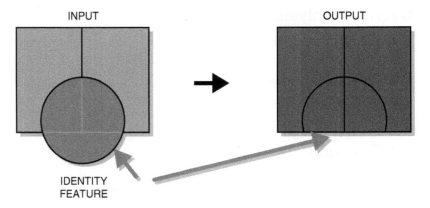

FIGURE 5-2 Identity function showing input and output rasters and the identity feature.

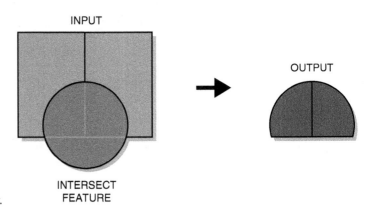

FIGURE 5-3 Intersection overlay.

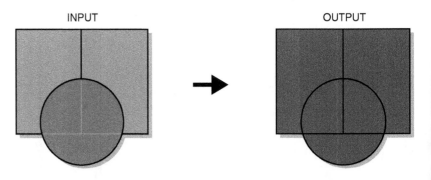

FIGURE 5-4 Union overlay.

These features act exactly the same as union and intersect Venn diagrams. With symmetrical difference, all or parts of features of the input and update features that don't overlap are created as the output feature class (Figures 5-5 and 5-6).

With raster overlay approaches, each grid cell of each grid layer references the same geographic location, making it well suited for combining characteristics of multiple layers into a single layer. Because each grid cell is assigned a number representing some characteristic, this allows you to employ any number of logical, mathematical, or statistical techniques to these

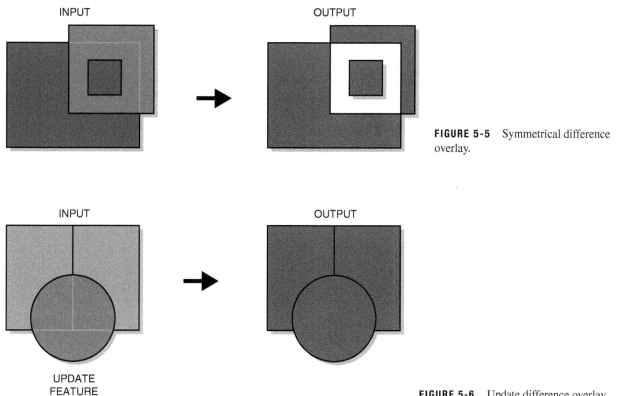

INPUT　　　　OUTPUT

FIGURE 5-5 Symmetrical difference overlay.

INPUT　　　　OUTPUT

UPDATE
FEATURE

FIGURE 5-6 Update difference overlay.

layers to derive new grid layers. Raster overlay tools include Zonal Statistics (located with the Zonal toolset), Combine approaches found in the Local toolset, Weighted Overlay in the Overlay toolset, and Weighted Sum found in the Overlay toolset as well (Figure 5-7). You will discover more about these possibilities in Chapter 6.

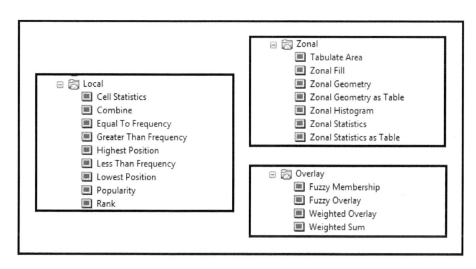

FIGURE 5-7 Local, zonal, and overlay tools available in ArcGIS.

ACTIVITY 5-1   OVERLAY ANALYSIS

This exercise asks you some basic questions about overlay analysis.

1. In your own words, describe what overlay analysis is.

2. Describe some problems that can be solved by using overlay analysis. Be specific, including general categories and what you expect to find.

3. Do a WebQuest and find some examples of the use of both vector and raster overlay and describe them, explaining what was overlaid and what resulted. Provide a name for the type of overlay analysis you found in each example.

4. Consider a series of complex land use polygon maps from different time periods (e.g., 1970, 1980, 1990, 2000, 2010) that you wish to overlay to evaluate how the land uses have changed through time. Assume that you digitized each map from available unrectified 1:24,000 aerial photography for each time period. Answer the following questions.

a. Do you think all the aerial photography would be identical in terms of accuracy? Why or why not?

b. Considering what you answered in question a, what does this suggest about the possible results of vector overlay?

c. So, given your answer to question b, would you suggest vector or raster overlay analysis? Specify the advantages and disadvantages you might expect from using raster over vector overlay in this case.

5. What is the purpose of a union overlay operation?

6. Describe what intersection overlay is.

7. Explain, using a concrete example, the purpose of symmetrical difference overlay.

8. Extra question: What is a spatial join? How does it work, and what does it show?

Proximity Analysis

Among the standard tools of all GIS software is its ability to measure distance. From this simple tool comes the advanced ability to perform proximity analysis, or the measurement of the nearness of one or more objects to others. Within that framework, one is able to answer questions such as these:

- How close is my house to the school my children will attend?

- Do highways designated as hazardous cargo routes pass within 1,000 meters of a river?

- How far is Los Angeles from San Francisco?

- What is the nearest grocery store to my home?

- What is the distance between a church in my buildings layer and a pond in the hydrology layer?

- What is the shortest patch along the road network from my house to your house?

As with Overlay tools, these proximity tools are divided into vector (feature based) and raster types. Feature-based proximity tools are used to identify proximity (distance) relationships. One of the more common such tools is the Buffer tool that creates area features at a user-specified distance (or several if you are using the multiple ring buffer variant) around existing input features (points, lines, or polygons) (Figure 5-8). These buffers can measure distance for polygonal lines in both inside and outside directions. They can be selected by attributes as well so that, for example, different sized streams (main versus primary or secondary tributaries, etc.) can get different sized buffers. It's important to note here that buffers are often used in conjunction with overlay operations in which the buffer polygons are overlaid on other features. For example, you might have a buffer around streams showing a 100-year flood zone (this would be based on an elevation layer as well as distance), and you would then overlay the buffer on existing houses to determine if homeowners were eligible for flood insurance.

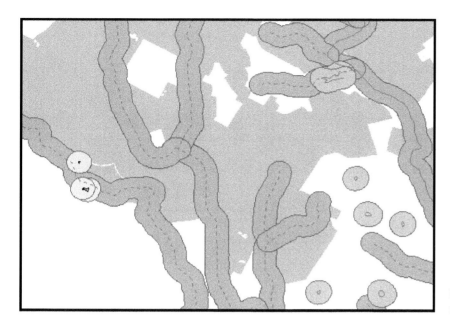

FIGURE 5-8 Buffers around existing features.

The **Near tool** calculates distances from features in one feature class to the nearest features in another feature class. This works for points, lines, and polygons. You could, for example, evaluate the nearness of summer cottages to a lake, distances from a homestead to an industrial polygon, or distance from a stream to a scenic drive. The output provides actual distances as well as feature identifiers and, if you wish, coordinates and angles to the nearest features. Additionally, you can create classes based on groups of distance measures for an easy-to-read output. A related tool, the Point Distance tool, looks at distances between two sets of places such as grocery stores and bus stops, allowing you to examine the relationships between food availability and transportation. You might also find Point Distance useful for locating water tanks in rangeland environments so you know where your cattle might have access to water.

Similar to the Near and Point Distance tools is Thiessen polygons. These polygons divide geographic space and allocate it to the nearest point feature. You can envision this as a 2-D version of a bubble pipe in which spheres become interconnected as tightly as possible. When surface data are not available, Thiessen polygons are often used to create a surface composed of proximal (the nearness of) polygons around the point features. You might do this, for example, to look at the spatial allocation of ant nests or prairie dog mounds when these species attempt to be as far away from each other as possible to avoid conflict. **Thiessen polygon analysis** has been proven useful for ecological studies when researchers wish to characterize the climatic variables from a sparse network of weather recording stations.

Distance tools are also available for Network databases from within the Network analyst extension (Chapter 7) and include an ability to solve routing problems, find the closest facilities, and determine service areas. Rather than measuring distance anywhere in a layer, such approaches are limited to the transportation network because it assumes you will not be traveling off road.

Raster-based distance tools are part of the Spatial Analyst extension. These tools can be grouped into Euclidean direction that gives a grid cell a value that indicates the direction of the nearest input feature, **Cost Distance** that adds a cost to the analysis of distance—costs being things like hazards, difficulty, energy expenditure, gas costs—that might vary based on factors like terrain and ground cover. The idea behind Cost Distance is to identify the least cost path from one place on a surface to another. Path distance that actually extends the Cost Distance tools allow the user to use a separate cost layer that includes things like the additional distance added to the horizontal distance added by topography in addition to the horizontal cost factor. Rather than producing a least-cost path, it creates a surface of path costs. Together, these Feature- and Raster-based tools provide a rather robust set of analytical techniques for working with distance and using it to determine a wide range of proximity measures.

ACTIVITY 5-2 | **PROXIMITY ANALYSIS**

The following simple exercises and questions review the concepts of proximity analysis as performed in both raster and vector tools.

1. In your own words, describe what proximity analysis is and provide some examples of how you could use it.

2. Describe the difference between Feature-based and Raster-based proximity analysis methods.

3. Perform a WebQuest and find examples of Buffer Analysis, Cost Path, and Path distance. Describe what the authors of these examples were trying to do.

ARCGIS LESSON 5-1 | BUFFERS

The purpose of this exercise is to give you an opportunity to experiment with the creation of buffers using ArcGIS. There are many forms of **Buffer Analysis**, but this particular exercise asks you only to do a simple buffer so you become familiar with the tool.

Steps

1. Open up ArcGIS and navigate again to Exercise 1 data in the ArcGIS tutorial dataset.

2. Turn off all the layers except BoundaryA (for reference) and HydroL. You are going to create a buffer around the HydroL layer.

3. From the Geoprocessing pulldown menu, select buffer.

4. When the buffer menu appears, select HydroL as the Input Features, Linear unit at 100 meters.

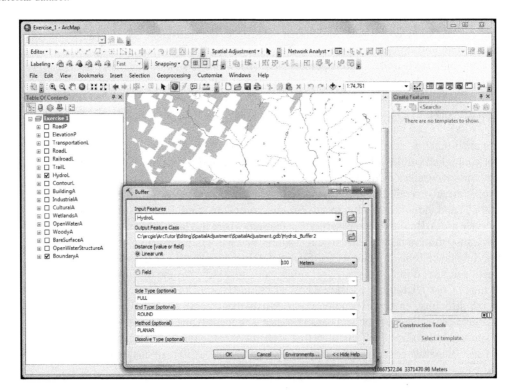

5. When you are set click OK. Be patient; this may take a minute or two to calculate. When you have finished, you should see the following.

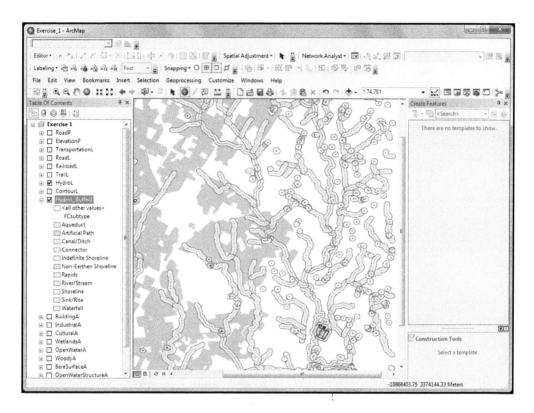

6. Notice that a new layer has been created.

7. Take a screenshot and turn it in to your instructor.

8. In a short report, explain why there are different categories of buffers in the legend and what they mean.

Summary Statistics

Maps contained in GIS databases by definition contain data on the characteristics of feature attributes as well as the specifics of their locations. While these data are contained in the database, they need to be compiled and analyzed statistically before they can truly be considered information—data in the context of a question. While some information might be observed from the map, the tables provide potential information about data trends (spatial), distributions of phenomena, or whether patterns might be present that are not immediately obvious from the map. In Chapter 4, you learned about how to identify and/or select individual features. **Statistical analysis**, on the other hand, is designed to provide you with information about the characteristics of the set of features and how they are characterized as a group. Although some advanced statistical techniques, notably the spatial statistics tools, are part of a geoprocessing toolkit, there are many spatial tools that are quite useful but do not have their own geoprocessing toolkit. Instead, the tools are part of ArcMaps basic toolbars and menus. Spatial statistics are rather advanced techniques and are not included in this text. Instead, you will learn about the basic spatial statistics commonly found in ArcGIS.

Among the primary features of statistical analysis in ArcGIS is the ability to explore your data in a search for highs and lows, distributions, and measures of central tendency. Beyond their inherent value, these characteristics also aid in the selection of map category ranges and in finding outliers and data errors as well as reclassifying maps. Statistical analysis is dependent on the type of data your layers may contain but can contain Minimum, Maximum, Range, Standard Deviation, Mean, and other basic statistics. To access these statistical tools for any layer you have active, click on the ArcToolbox icon, click on the Statistics box, and three items, Frequency, Summary Statistics, and Tabulate Intersection appear (Figure 5-9). Frequency provides you with the number of objects you have selected. Tabulate Intersection is a cross tabulation of the area, length, or out of intersecting features between two feature classes. For the time being, you should focus on the Summary Statistics because it will provide you with a table of the statistics you request for the layers you selected (Figure 5-10).

FIGURE 5-9   Location of the Summary Statistics tool.

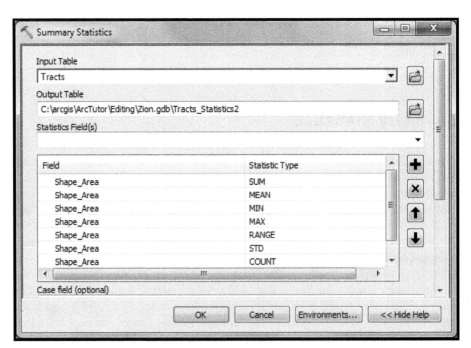

FIGURE 5-10   Summary statistics menu showing statistics you requested.

The following simple exercises and questions review the concepts of using Summary Statistics.

1. In your own words, describe why you might want to use Summary Statistics.

2. Describe how you would find the Summary Statistics tools.

3. Perform a WebQuest search to see how others have used Summary Statistics in their work. Provide a brief description of what they did and why they used the technique.

Table Management

Nearly all attribute data and, in many cases, entity data are stored as database tables. Feature classes, for example are tables that contain shape attribute columns, and rasters in Spatial Analyst can be viewed as tables of attributes. Typically GIS databases have tables containing attributes that are specifically related to a particular topic that can then be related to other tables by a common column of attributes. Prior to any analysis, you will spend a fair amount of time working with the tabular data so that they are set up properly for subsequent analysis.

Beyond acting as storage and search conduits, the tables themselves provide a few nice features that you can use to build new data. For example, if you open up a table in ArcGIS, you will notice that by right clicking on a column, you immediately have access to sorting capabilities as well as some of the same basic statistical techniques you saw in the last section (Figure 5-11). The Properties button allows you to pull up a menu that displays the field properties (Figure 5-12). As with any spreadsheet, the tables also allow you to turn a field off or freeze a column so you can effectively navigate your tabular data. All of these basic features make it easy for you to scroll around in your tables so you know what is there and so that you can make

FIGURE 5-11 Sorting capabilities dialog box found by right clicking a column in the table.

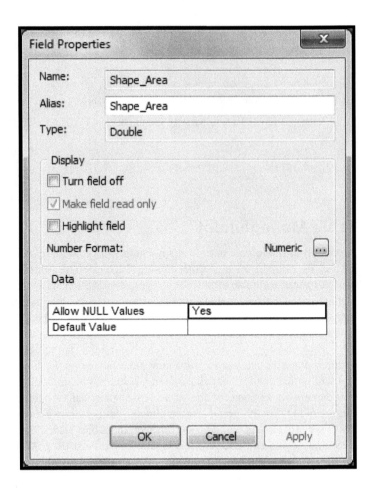

FIGURE 5-12 Properties of the selected field.

decisions about how to manage the data either for subsequent analysis or for map output (e.g., category selection).

There are times when you need to add new layers of data to your database. If you happen to have layer data, you can just add it to your existing layers. But what if you have a set of tabular data that are associated with a set of existing polygons in your database but are not geocoded? One of the nice features of the tables inside ArcGIS is that there is an Add Field tool that allows you to create the attribute field you wish at any time. Just by clicking the Table Options button on the top left of your table and selecting it from the available options (Figure 5-13), you can add a column of data to your database. You will also notice a large array of additional tools available to you including those allowing you to perform selections of attributes (see Chapter 4 on map queries), join and relate this table to other tables you might have in your database, generate reports, add your tabular data to a map layout, create graphs, print, and many more.

One of the most powerful tools in the table management toolkit is called the **Field Calculator**. The Field Calculator allows you to create or update the values of a field (column) by creating a mathematical equation that is applied to the data (Figure 5-14). An example might be

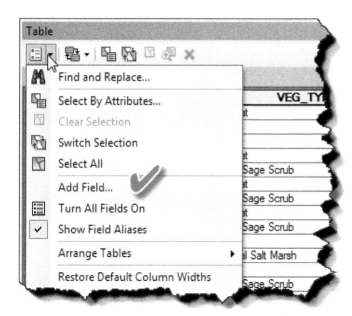

FIGURE 5-13   Add fields found in the table options.

OBJECTID *	Shape *	HOLL		PE	Shape_Length	Sha
1	Polygon	11300	Sort Ascending		353.924583	6
2	Polygon	13111	Sort Descending		10155.121101	1199
3	Polygon	13130	Advanced Sorting...		2380.510623	381
4	Polygon	11300			3670.16697	30
5	Polygon	32500	Summarize...		2498.556584	134
6	Polygon	11300	Σ Statistics...		1424.97827	64
7	Polygon	32500			3117.744081	352
8	Polygon	11300	Field Calculator...		2679.583267	98
9	Polygon	32500	Calculate Geometry...		583.317816	11
10	Polygon	37000			1410.429085	115
11	Polygon	52120	Turn Field Off		3466.738265	231
12	Polygon	37000			8292.402989	802
13	Polygon	32500	Freeze/Unfreez		322.589703	
14	Polygon	42110			95.382157	
15	Polygon	42200	✕ Delete Field		203.372398	2
16	Polygon	11300	☞ Properties...		4200.268148	330
17	Polygon	32500			6452.066354	622
18	Polygon	32500		Diegan Coastal	2864.644375	500
19	Polygon	42000		Valley and Foot	1629.551255	87
20	Polygon	63300		Southern Riparian Scrub	1348.433643	65
21	Polygon	32500		Diegan Coastal Sage Scrub	15862.601568	299
22	Polygon	11300		Disturbed Habitat	3120.204939	228
23	Polygon	11300		Disturbed Habitat	3517.955979	354
24	Polygon	11300		Disturbed Habitat	3878.111654	345
25	Polygon	13130		Estuarine	6488.004272	33
26	Polygon	52120		Southern Coastal Salt Marsh	3554.275191	161
27	Polygon	13130		Estuarine	2094.644496	96
28	Polygon	13400		Beach	3610.045095	179
29	Polygon	52120		Southern Coastal Salt Marsh	7185.931179	89
30	Polygon	32500		Diegan Coastal Sage Scrub	3050.199968	278
31	Polygon	11300		Disturbed Habitat	1211.901439	85

Field Calculator

Populate or update the values of this field by specifying a calculation expression. If any of the records in the table are currently selected, only the values of the selected records will be calculated.

Vegetation Type

I◄ ◄   0 ► ►I   ☷ ☴ (0 out of 2333 Selected)

FIGURE 5-14   Location of the Field Calculator.

that you have a table with the population by county for your state. You want to map these population data using a choropleth (polygon-by-polygon) mapping approach, but you are aware that you shouldn't be using raw data for such an exercise. Instead, you want to update the field so that it is a percentage of the entire state's population. Your formula would simply be that each cell in the column would be divided by the total state population—to obtain the fraction and then multiply it by 100—to get the percentage. This would be much the same as performing the operation in any spreadsheet. It is usually best that you not do this with the original field but create a blank field and perform the calculations on it. That way you preserve the integrity of your original database as well as having a new field with which to work.

Just below the Field Calculator is the Calculate Geometry tool that, as with the Field Calculator, allows you to recalculate or create geometric values in a column derived from existing geometric values such as area, perimeter, length, and so on. You might use this approach, for example, to calculate the actual, on-the-ground, length of roads. Before you use the calculator, you also need to make sure that all layers are in the same projection and have the same datum and zone by right clicking on the layer and going to Properties. This ensures that your map layers will all be aligned. If they aren't, you need to go to the catalog and change the coordinate system's properties so they match. Then you begin as before by creating a new field with appropriate properties. At the top of that new field, you right click, select Calculate Geometry, select the Geometric Property (e.g., length or area), match the coordinate system, define the units of measure, and click OK. You are finished.

This approach to calculating new fields is faster than doing it within an edit session because the computer doesn't have to operate on the graphics at all while the calculations are being performed. A word of caution, however, is that once you do this, the operations cannot be undone as it can when using the graphical user interface. So if you make a mistake, especially if you do the calculation on the original field rather than on a newly created field, you could very well mess up your database. This is a good place to suggest that you keep copies of your databases so you don't have to worry about such mistakes.

ACTIVITY 5-4 TABLE MANAGEMENT

The following simple exercises and questions review the concepts of using table management.

1. In your own words, describe some basic table management capabilities of ArcGIS. Next to each, indicate why you would use it and/or why it is important.

2. Describe in your own words how to use the Field Calculator.

3. In your own words, describe how to create a new field and why you might do this.

4. What cautionary note did your author provide regarding using the Field Calculator or the Geometry Calculator within the tables as opposed to using the graphical user interface of ArcGIS?

ADDITIONAL READING AND RESOURCES

DeMers, Michael. *GIS for Dummies*. Indianapolis, IN: Wiley, 2009.

Gorr, Wilpen L., and Kristen S. Kurland. *GIS Tutorial 1: Basic Workbook*. Redlands, CA: Esri Press, 2010.

KEY TERMS

Buffer Analysis: An operation in which a preselected distance is measured from any point, line, or polygon feature and whose output is a set of polygons.

Cost Distance: Measurement of the cost associated with travel other than merely distance.

feature overlay: A collection of overlay methods involving vector points, lines, and polygons.

Field Calculator: A tool for creating a new field based on calculations of one or more other fields in the table.

Near tool: An ArcGIS tool that compares the distances of one set of features to another.

overlay analysis: A category of spatial analytical techniques involving the mathematical comparison of one data layer with another.

proximity: Nearness of one geographic feature to another.

raster overlay: A collection of overlay methods involving the raster data layers in a GIS.

statistical analysis: Any of a number of techniques for summarizing or characterizing tabulated data.

Thiessen polygon analysis: A method of determining proximity involving the allocation of geographic space around point objects.

CHAPTER 6

Raster Operations

LEARNING OBJECTIVES

Here is the content you will learn in this chapter:

1. List and describe the basic properties of raster datasets.
2. Define Map Algebra and explain its rules.
3. Locate and use the Spatial Analyst module.
4. List and define the Spatial Analyst terminology.
5. Become familiar with the Spatial Analyst classes.
6. Through application, become familiar with the Image mosaic.
7. Use the Image mosaic toolset.
8. Through application, be able to identify and describe the Spatial Analyst toolsets.
9. Through application, be able to demonstrate and describe the analysis environment in Spatial Analyst.
10. Use Spatial Analyst to perform a raster analysis.

BEHAVIORAL INDICATORS

When you are finished with this chapter, you will be able to:

1. List and describe basic raster dataset properties.
2. Explain in simple terms what Map Algebra is and its basic rules.
3. Open and become familiar with the Spatial Analyst module.
4. List and define Spatial Analyst terms.
5. Briefly describe Spatial Analyst classes.
6. Define and describe what an image mosaic is.
7. Describe the use of image mosaicking.
8. List and describe the toolsets of Spatial Analyst.
9. Briefly describe the analysis environment of Spatial Analyst.
10. Perform a simple raster analysis using Spatial Analyst.

Chapter Overview

This chapter covers arguably the most powerful modeling and analysis tools in the ArcGIS arsenal. Because there is so much to cover, there will be fewer hands-on sections in this chapter simply because to become good at these tools, you need to know the underlying principles of raster GIS analysis. While vector seems more accurate visually, its analytics are restricted by the available data structures whereas raster analysis is extremely robust but sometimes can lack the visual appeal of vector.

The chapter begins with a refresher of raster data and how they represent geographic features. It then proceeds to consider the **Map Algebra** language that permits the huge variety of methods and procedures for analytical work. Next, the chapter covers the basics of the Spatial Analysis module—its interface, terminology, and classes. Because digital satellite imagery is so prevalent in the raster domain, the chapter examines the image classification toolset, its terms, and how it can be implemented inside the Spatial Analyst toolkit.

In preparation of the final section that allows you to perform some analysis with Spatial Analyst, the chapter examines the wide variety of toolsets from which to choose. Each is explained briefly, and then the chapter covers the overall analysis environment available to leverage these tools in Spatial Analyst. The exercise in the final section provides you with an opportunity to use the Spatial Analyst module but hardly touches the surface of its power. You might want to consider some additional reading or coursework to enhance your Spatial Analyst skills if you wish to be a robust modeler.

Raster Data and Map Algebra

A necessary first step to understanding the ArcGIS Spatial Analyst module and the Map Algebra language that drives it is to recall how raster data work. Recall from the Chapter 2, Raster Data Model section that raster data divide the earth into discrete units called *grid cells*. Usually square or parallelograms (in the case of satellite imagery units called *pixels*), each covers a small portion of the Earth. While these cells are linked to the geographic grid, for the sake of simplicity, you can assume that the cells rest on a flat surface.

Raster datasets contain information on the following properties: data source, raster information, extent, spatial reference, statistics, geodata transform, raster metadata, and wavelength. The data source describes the name, type, and location of the data and includes database server information for raster data on a database server. If the raster data are derived products (not sourced), there will be no data source information displayed.

Raster information includes the number of rows and columns, number of bands (imagery), cell size (x,y), uncompressed size, format, source type, pixel type (signed, unsigned, inter/rational), pixel depth (1 through 64), NoData value, colormap (present/absent), pyramids, compression type, and many more. Extent is the rectangular boundary containing all the raster dataset's data. Spatial reference is the dataset's coordinate system. Statistics indicate minimum and maximum, mean, standard deviation, and number of raster classes. Geodata transform describes the mathematical model that geometrically transforms the pixels. Finally, key metadata is associated with a raster product including things like sensor name, product name, and acquisition date. While all this seems a bit trivial, it will come in handy when you need to model using these data.

As the grid cells for multiple layers are put into the database, each layer must exactly correspond in x and y coordinates to the other layer. This is an indelicate way of saying the cells must be coregistered. It is not only important from a geographic perspective that these cells be coregistered but also from a calculation perspective. Each layer's grid cells must line up, stacked right on top of each other, so that they can interact with each other on a one-on-one basis.

As in most computer-based systems, there is a language that allows the user to communicate with the algorithms. The same is true of the ArcGIS Spatial Analyst module that operates on a language called Map Algebra, first developed in the early 1980s at the Yale School of Forestry. This language is a simplification of matrix algebra in which the numbers associated with individual grid cells are linked to the grid cell positions. The language allows for the use of a wide variety of statistical, mathematical, logical, spatial analytical, or Boolean operators to combine raster map data through their selective use. As an example, two simple matrices of 2×2 could be added thus:

$$\begin{pmatrix} 4 & 5 \\ 7 & 8 \end{pmatrix} + \begin{pmatrix} 6 & 10 \\ 9 & 11 \end{pmatrix} = \begin{pmatrix} 10 & 15 \\ 16 & 19 \end{pmatrix}$$

Instead of thinking of these as numerical matrices, think of them as small, four-cell raster maps like that in Figure 6-1. Now you see a simple Map Algebra expression in the form of a graphic. Verbally, this would be map 1 plus map 2 equals map 3.

This same process applies to all the mathematical, statistical, spatial analytical, and Boolean operators and, unlike matrix algebra, all operations are linked to the position of the grid cells. So, if you are multiplying these two matrices, you would multiply 6×10, 10×15, 9×16, and 11×19, which is different than the process in matrix algebra. In its original incarnation, Map Algebra required that statements such as that in Figure 6-1 (bottom) were typed in what was called "command line" entry. Today, fortunately, ArcGIS's Spatial Analyst toolkit allows you to access the toolkit using the Python programming language, integrated development environment (IDE), the Python Window inside ArcGIS, and the Raster Calculator tool. All of these are part of the Spatial Analyst vocabulary and user interface.

The Map Algebra that drives Spatial Analyst and many other raster GIS software packages consists of a basic set of rules exemplified by the last paragraph. The general structure (**syntax**) of Map Algebra is as an equation in which the assignment operator ($=$) separates the name of the raster output on the left and the action required to obtain that output on the right. So, in equation form, the structure is Output_Name$=$Action. These expressions are composed of five components.

1. Geoprocessing tools.

2. Functions (e.g., reclassify, sin, orient) used in Map Algebra expressions.

3. Operators, which are symbols used to represent the mathematical operations (e.g., $+$, $-$, $/$).

4. Input elements including variables, objects, constants, numbers, features, and—oh yes—rasters. (Note here that Map Algebra is not limited to just the raster layers.)

5. Output data, which is a raster object referencing temporary raster data (e.g., Outras . . . meaning output raster). Simply put, a raster object is typically a raster dataset.

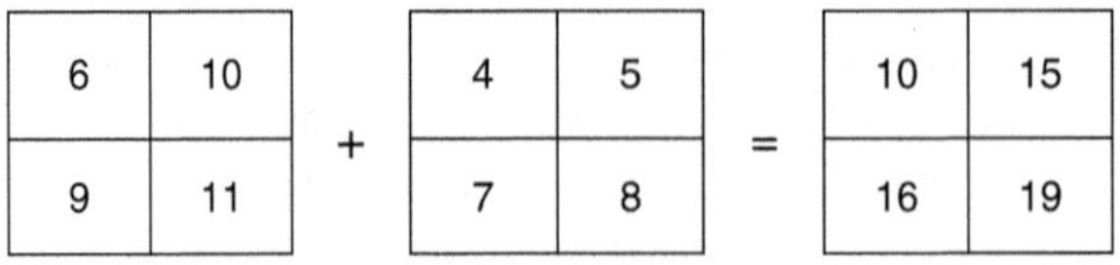

6	10
9	11

$+$

4	5
7	8

$=$

10	15
16	19

FIGURE 6-1 Four-cell raster maps showing the relationship between matrix algebra and Map Algebra.

ACTIVITY 6-1   **RASTER DATA AND MAP ALGEBRA**

In the following activities, you will get a chance to refresh your memory about what you read regarding raster data and Map Algebra.

1. List and describe at least five raster data properties. When you have finished, explain why you need to know this.

2. Search the ArcGIS help menu and the Internet to gather what is known about Map Algebra. Having done so, define it in your own words to demonstrate your mastery of the concept.

3. What are the basic rules of Map Algebra? In answering this, describe the five components of a Map Algebra statement. *Note:* It is important to learn this because it is the basis of all Spatial Analyst operations.

Spatial Analyst Module, Terminology, and Classes

Spatial Analyst is an extension to ArcGIS that needs to be loaded unless you loaded the module when you first turned ArcGIS on. If you have done so, all you have to do is select the Customize pulldown menu and select Extensions (Figure 6-2). Check the box next to the Spatial Analyst tool. After you have done that, you will notice that when you go to the toolbox menu, Spatial Analyst is now available for you to use (Figure 6-3).

Spatial Analyst is not a single tool. Instead, it is a rather robust language and set of tools that allow a wide variety of analytics. This is the terminology of ArcGIS that you must master to be able to take advantage of its full power. Figure 6-4 shows you the tools available as of version 10.3.

Each is designed for a specialized set of analyses, such as Groundwater tools used by groundwater hydrologists and hydrogeologists and Hydrology tools for surface water analysis. While these all conform to the rules and syntax of Map Algebra, among the most powerful of these tools is the Map Algebra tool itself (Figure 6-5).

The Map Algebra tool is one of many embedded in the Spatial Analyst toolkit which is quite robust, including at least 170 tools and 22 toolsets to build models and perform spatial analysis on raster data.

Within the Spatial Analyst toolkit, there is a module that extends the capabilities of Spatial Analyst through a series of classes used to define parameters for the tools themselves when those

FIGURE 6-2 Spatial Analyst checkbox.

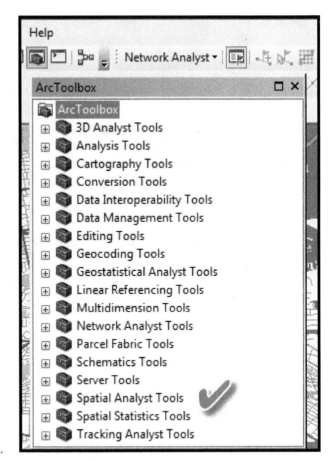

FIGURE 6-3 Location of Spatial Analyst tools in the ArcToolbox menu.

tools have various numbers of arguments. These parameters may vary depending on the parameter time (e.g., neighborhood type) or when the number of entries varies based on the specific situation (e.g., a table used to reclassify spatial attributes, especially in raster). The classes provide you with an enormous amount of flexibility to change the individual entries in the parameter, thus giving you an amazing power to produce analysis. Following is a list of the classes and their definitions:

- **Fuzzy Membership.** Set-based membership function for fuzzy logic calculations

- **Horizontal Factor.** Horizontal factor for Path Distance tools

- **Kriging Tools.** Used for setting the models for Kriging-based interpolation modeling

- **Neighborhood.** Defines the input neighborhood parameters for a series of tools

- **Overlay.** Creates the input tables for Weighted Sum and Weighted Overlay tools

- **Radius.** Defines the radius for the Inverse Distance Weighted and Kriging interpolations

- **Remap.** Defines the tables used for reclassification

FIGURE 6-4   The list of available Spatial Analyst Tools in ArcGIS.

- **Time.** Identifies the time interval to use in the solar radiation tools

- **Topo Input.** For use in defining the input to the Topo to Raster tool

- **Transformation Functions.** Define the transformation functions for the Rescale By Function tool

- **Vertical Factor.** Defines the vertical factor for the Path Distance tool

Just so you know, you are not likely to experience many of these options in your first course in GIS, but it's nice to know they are there when you do encounter them. This also demonstrates the enormous, and often underused, raster tools available in today's modern GIS software. Simply being aware of the tools is often enough to seek them out when you want to do complex modeling. Each of the classes in the list has numerous options, making its ability to model complexity even more impressive.

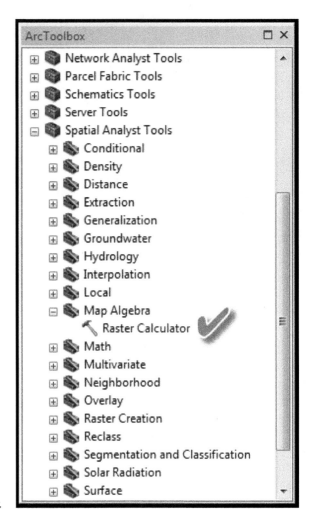

FIGURE 6-5 Location of Raster Calculator.

One of the strengths of Map Algebra and the Spatial Analyst Module is its advanced ability to create robust expressions in its algebralike language. The syntax of these expressions is in the format of English-language sentences. To make expression creation easier, ArcGIS has created an efficient tool in a calculatorlike interface, the Raster Calculator (Figure 6-6).

The Raster Calculator contains a window for you to select Layers and Variables, a set of buttons called Operator Buttons, another window for Tools on the right, and on the bottom, another window that displays the Expression. On the bottom is the path to the Output raster that will result from the operation created in the expression. The available mathematical operators include division, equal to, not equal to, Boolean (And), multiplication, greater than, greater than or equal to, Boolean (Or), subtraction, negate, less than, less than or equal to, Boolean (XOr) addition, and Boolean Not. These are combined with a wide array of commands for conditional selection, mathematical, and statistical analysis. Although there are some minor differences, the Raster Calculator tool works well with Model Builder to create advanced models. Proceed to Activity 6-2 to see how much of this information you recall.

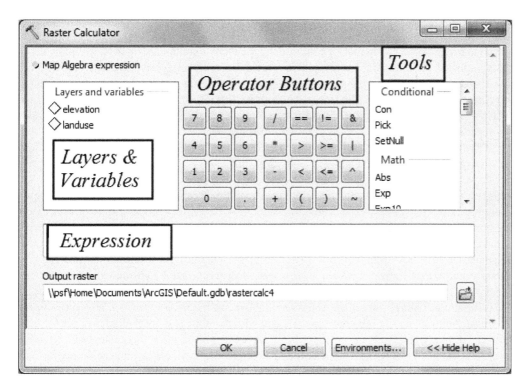

FIGURE 6-6 Raster Calculator with parts shown.

ACTIVITY 6-2   SPATIAL ANALYST ENVIRONMENT AND CLASSES

In the following activities, you will have a chance to refresh your memory about what you read regarding the Spatial Analyst environment and the classes of variable with which you can work.

1. In your own words, describe what the Spatial Analyst extension is and what it does.

2. List the classes available for Spatial Analyst. Describe generally what the purpose of these classes is.

3. Pull up an image of the Raster Calculator on your screen and take a screenshot. Trim away material not included on the actual Raster Calculator. Then label the diagram showing the different windows and what they do.

4. Perform a Google WebQuest for the words Raster Calculator, and select an image as your output. From the graphics

you observe, pick three images that show Map Algebra expressions. Select them, click on the webpage associated with them, write the expression down, and describe what the expression is designed to analyze.

5. Generally, Spatial Analyst answers four complex question types. Seek out the documentation for spatial analysis from the Esri website and, in your own words, provide a description of each of those in the following the list:

   a. Find suitable locations

   b. Manage risk

   c. Visualize patterns

   d. Model cost

Image Analysis Toolsets and Terms

Although raster data come in many forms, among the more common are images, including aerial photographs, U.S. Geological Survey quadrangles, and satellite imagery. Although ArcGIS is not considered an image-processing package, it possesses a wide range of tools for display, analysis, and classification of imagery. The **Image Analysis toolkit** is found in a dockable window. On the main menu, click on the Windows pulldown and select Image Analysis. This window provides five separate components: Layer List, Options, Display, Processing, and Mensuration.

The Layer List provides a list of all the raster layers that are active in the data frame including such items as the **Mosaic Dataset** image layer, image service, and the WCS (World Coordinate Systems) layers (Figure 6-7). Mirroring the table of contents, window allows you access to the layer properties dialog box and gives you the capability to remove layers. Unlike the Table Of Contents, it does not allow you to reorder the layers. One important property of the Layer List is that it allows you to "accelerate" a selected layer. This process increases a layer's rendering performance. In this mode, you can pan and zoom around an image, allowing the computer to take advantage of advanced graphics cards. The use of the graphics cards is optional and must be activated under the ArcMap options. A nice feature of the Layer List window is that you can select more than one layer at a time for either display or processing.

The Image Options button is located at the top of the Image Analysis Window. It opens the Image Analysis dialog box that provides you with the ability to set the defaults for some of the tools in that window. Many of these tools are used primarily for image analysis, which is beyond the scope of this book. The other toolkits include display tools to enhance the appearance of the imagery, processing that provides one-click access to common image processing techniques including orthorectification, clipping, and masking. Finally, the mensuration panel contains tools for measuring point, distance, angle, height, perimeter, and areas from an image or Mosaic Dataset.

The Mosaic Dataset is a really nice feature because it allows the user to manipulate large sets of images or other raster data easily. It creates sets of tiles that allow the user to select specific parts of the database. The other tasks described here rely heavily on the Mosaic Dataset. In ArcGIS Lesson 6-1, you will get a chance to learn how this works.

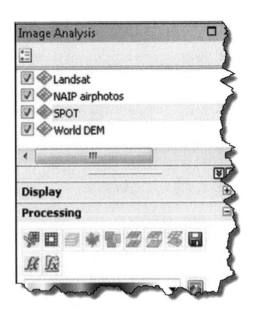

FIGURE 6-7 Available raster datasets for your exercise. In this case, they are Landsat data, NAIP airphotos, SPOT, and the World DEM.

With the number of digital satellites providing land-based data, there is a similar increase in the available image data for use in GIS. These uses could include enhancing the imagery to allow the user to see things more clearly, and for providing the ability to see things that could not otherwise be seen by teasing out the detail or even hiding the detail to provide a clearer overall picture without the distractions from the detail. The general term for these techniques, *image processing*, is a discipline all its own, but ArcGIS now contains many of the basic image-processing operations, thus not requiring you to become familiar with two completely different packages. To prepare you for the use of ArcGIS for this activity, please answer the following questions. Use the documentation of your software and/or the ArcGIS tutorials and documentation found online.

1. Provide a list of at least six different sources of raster image data. Next to each, provide a quick description of what each might be used for.

2. What does the term *multispectral* mean, and why is it important in image analysis?

3. Pull up the Image Analysis extension of ArcGIS and take a screenshot that shows where the tool is and how to locate it. Use arrows and/or other means to show different parts of the user interface.

4. Some GIS analysts say that the Image Analysis allows the user to interrogate the images and the individual pixel (image raster) values. Research this via the Internet and provide a concrete example of this process that explains not only what it does but also why it is useful.

5. Use the Internet to describe what a **GeoTIFF** image is. How does it differ from a traditional TIFF image?

ARCGIS LESSON 6-1 | CREATING A MOSAIC DATASET

This lesson gives you the opportunity to create a Mosaic Dataset containing GeoTIFF raster dataset files using the geoprocessing framework of ArcMAP.

Part I. Mosaic Creation

1. Start ArcMap.
2. Click Cancel on the ArcMap—Getting Started window.
3. Click the Catalog button 🔲 on the Standard toolbar. This will open the Catalog window.
4. In the Location text box, type C:\arcgis\ArcTutor\Raster and press ENTER. This location is added to the Catalog Tree under the Folders Connection heading. *Note:* If you placed your tutorial data elsewhere, navigate there instead.
5. Right-click the Raster folder and click New > Folder.
6. Name the folder Exercises.
7. Right-click the Exercises folder and click New > File Geodatabase.

8. Rename the new file geodatabase Image.gdb.

   Each map has a default geodatabase that it uses. This is the home location for the spatial content of your map and is used for adding datasets and saving the results of editing and geoprocessing.

9. Right-click the ImageGDB geodatabase in the Catalog window and click Make Default Geodatabase.

   Now you are going to create a new Mosaic Dataset.

10. Right-click ImageGDB in the Catalog window and click New > Mosaic Dataset. This will open the Create Mosaic Dataset tool dialog box.

11. Type Amberg in the Mosaic Dataset Name box.

12. Click the Coordinate System browse button 🗒.

13. Expand Projected Coordinate Systems > National Grids > Germany, choose Germany Zone 4, and then click OK.

(*continued*)

ARCGIS LESSON 6-1 | (CONTINUED)

14. Click OK on the Create Mosaic Dataset dialog box. The progress bar will display the status of the active tool. Once the process is finished, a pop-up message appears.

     The Amberg Mosaic Dataset is created in the geodatabase and added to the ArcMap Table Of Contents. This is an empty Mosaic Dataset. You will be adding some raster data to it in the next steps.

When you add the Mosaic Dataset to the Table Of Contents, ArcGIS adds the dataset as a mosaic layer, which is essentially a special group layer. The top level has the name of the mosaic dataset (Amberg). There are also empty Boundary, Footprint, and Image layers.

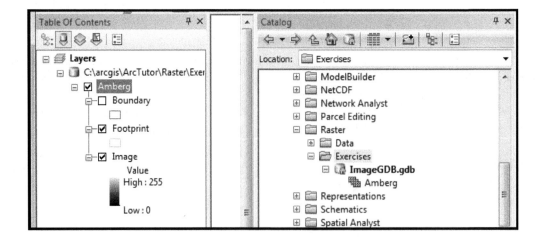

Now you're going to add rasters to the mosaic dataset.

15. Right-click the Amberg Mosaic Dataset in the Catalog window and click Add Rasters. This opens the Add Rasters To Mosaic Dataset tool.

16. In the Raster Type list, choose Raster Dataset.

17. Click the drop-down arrow and click Workspace.

18. Click the Input browse button.

19. Navigate to C:\arcgis\ArcTutor\Raster\Data\Amberg_tif and click Add.

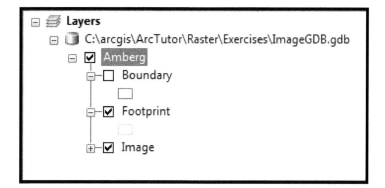

20. Check Update Overviews.

21. Click OK to run the tool.

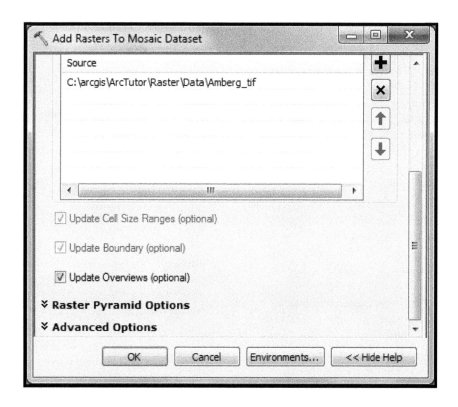

(continued)

ARCGIS LESSON 6-1 | (CONTINUED)

While the tool is running, the Mosaic Dataset cannot be edited; therefore, Lock Symbols appears over its layer in the Table Of Contents.

The progress bar will display the status of the operation as it is running. Once the process finishes, a pop-up message will appear, and the raster datasets will be added to the Mosaic Dataset. The footprints (lines) are created for each raster dataset, and the boundary is generated for the entire mosaic. The overviews will then be generated for the entire mosaic.

You might need to click the Full Extent button 🌐 to view the Mosaic Dataset.

Part II: Mosaic Property Modification

Mosaic Dataset properties can be modified. These properties affect how the mosaic image appears to the user and how the user interacts with the published dataset.

The next steps allow you to change the method of compression for the dataset and set the allowable methods of mosaicking. The compression method can affect the transmission speed. It is good practice to set a compression method rather than uncompressed to transmit the mosaicked image more quickly. You might want to publish the Mosaic Dataset as an image service for others to use, allowing clients to modify the setting to decompress the mosaicked image if they wish. The mosaic method defines the order in which the rasters are mosaicked together to create the image. You can choose one or more allowable mosaic methods and the one that will be the default. The user is able to choose from the methods you select.

1. Right-click the Mosaic Dataset in the Table Of Contents and click Remove. When editing the properties of a Mosaic Dataset that is open in the application, not all property changes are updated. You need to remove the Mosaic Dataset and add it back in.

2. Right-click the Amberg Mosaic Dataset in the Table Of Contents and click Remove. When editing the properties of a Mosaic Dataset that is open in the application, not all property changes are updated. You need to remove the Mosaic Dataset and add it back in.

3. Right-click the Amberg Mosaic Dataset in the Catalog window and click Properties. This opens the Mosaic Dataset Properties dialog box.

4. Click the Defaults tab.

5. Click the Allowed Compression Methods ellipsis button ⬚

6. Click the Default Method arrow and click JPEG.

7. Click OK.

8. Click the Allowed Compression Methods ellipsis button ⬚

9. Uncheck Closest to Viewpoint and Seamline. You are turning off Closest to Viewpoint because the Mosaic Dataset will not be used that way. You are turning off Seamline because you are not creating any. Thus, this method cannot be applied.

10. Click OK.

11. Click OK to close the Mosaic Dataset Properties dialog box.

Part III: Adding Metadata

Briefly defined, *metadata* is data about data. It describes the properties of the dataset so another user can understand how to use the data properly.

Steps

1. In the Catalog window, right-click the Amberg Mosaic and click Item Description.

2. Click the Edit button on the top of the window.

3. In the Title text box, type Amberg.

4. In the Tags text box, type orthophotos, Germany, city.

5. In the Summary text box, type A Mosaic Dataset containing several orthophoto TIFF images of Amberg, Germany.

6. Click the Save button at the top of the window.

7. Close the Item Description—Amberg window. You have completed creating the Mosaic Dataset and defining metadata.

Part IV: Exploring the Mosaicked Image

Now it's time to explore what you've created.

1. Drag and drop the Amberg Mosaic Dataset from the Catalog window into the display.

2. Use the tools on the Tools toolbar to pan and zoom around the mosaicked image.

3. Right-click the Image layer in the Table Of Contents and click Properties. The Layer Properties dialog box for the mosaicked image is opened. This is similar to the dialog box for any other raster layer.

4. Click the Status tab.

Here you get a chance to explore the properties of the mosaicked image (e.g., the number of rows and columns and the transmitted size).

5. Note the value for Transmitted Size.

6. Click the Display tab.

You can modify the compression method from JPEG that you set it to earlier to something else or change the quality value.

7. Click the Transmission Compression arrow and click None.

8. Click Apply.

9. Click the Status tab. The transmitted size has increased, which means a larger mosaicked image is being displayed.

10. Click on the Mosaic tab.

11. Click the Mosaic Method arrow, and then click Closest to Center.

12. Click OK to close the Layer Properties dialog box.

13. Pan and zoom around your image. Notice how the images are ordering themselves differently due to the changed mosaic method.

14. Click the Select Features By Rectangle button. You will see rectangles similar to these.

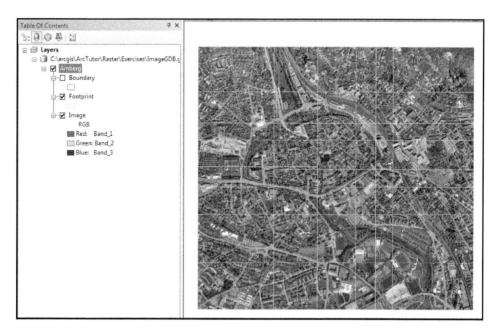

15. Click the middle left of the mosaicked image. Notice a number of footprints selected (highlighted lines).

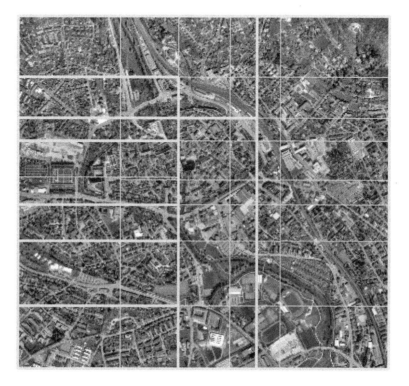

(*continued*)

ARCGIS LESSON 6-1 | (CONTINUED)

16. In the Table Of Contents, right-click Footprint > Open Attribute Table.

17. Click Show Selected Records.

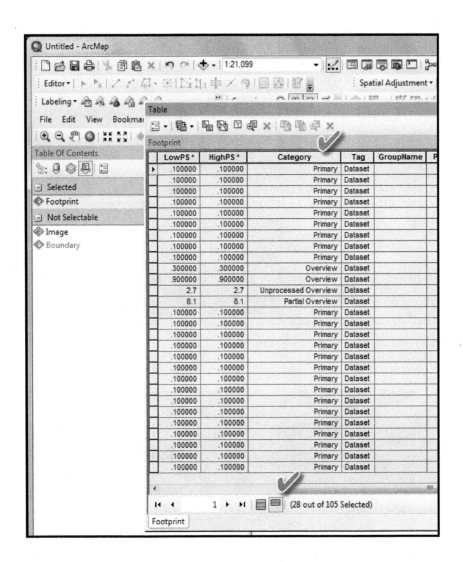

18. Scroll across the table to the Category field.

 You should see Primary and Overview within the field. The rows with Primary values are the orthophotos that you added. The rows with Overview values are the overviews that the system generated for the Mosaic Database when you added the orthophotos to the Mosaic Dataset. The overviews allow you to view the mosaicked image at all scales.

19. In the Table Of Contents, right-click Footprint > Selection > Reselect Only Primary Rasters.

 You will see the primary rasters selected in the table.

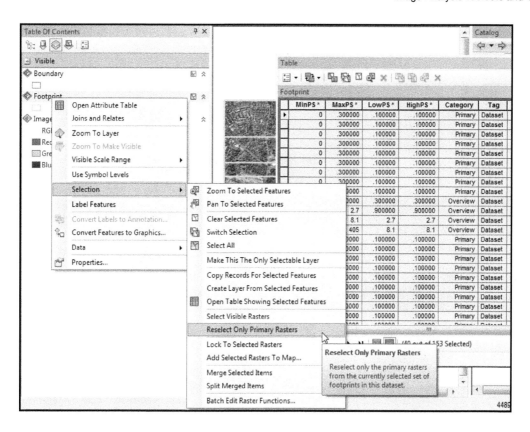

20. Close the Table window.

21. Right-click Footprint > Selection > Add Selected Rasters To Map.

22. In the Group layer name text box, type Primary.

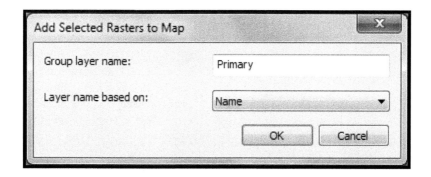

23. Click OK. The selected items in the Mosaic Dataset will be added to the Table Of Contents as individual layers with the Primary group layer.

(*continued*)

ARCGIS LESSON 6-1 | (CONTINUED)

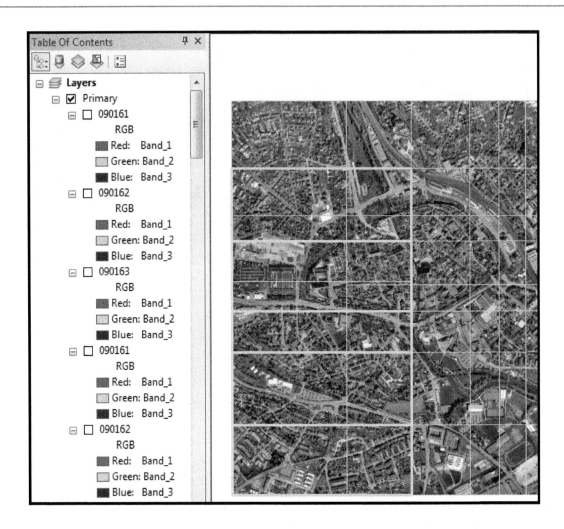

24. Uncheck the Amberg layer, check on the newly added layers, and explore these layers.

25. Right-click one of the layers contained with the Primary group and click Properties. You can see from the tabs that this is still a Mosaic Dataset layer.

    Take a screen capture of your work here to turn in to your instructor.

26. Click the Mosaic tab.

    You should see the Mosaic method has been changed to Lock Raster and there is a Lock Raster ID number. If you open the properties for the other layer, you should see that the Lock Raster ID number is different.

This is just a single example of how the Mosaic Dataset can be used as a way to provide a single mosaicked image while providing access to the individual items.

Take a screen capture of your work to turn in to your instructor.

27. Close ArcMap.

    The Mosaic Dataset you have created is now ready to publish as an image service using ArcGIS for Server. You can also use the Mosaic Dataset as a layer within ArcMap or ArcGlobe.

ARCGIS LESSON 6-2 | SETTING FEATURE TEMPLATE PROPERTIES

This lesson introduces you to some of the basics of prepping and using Spatial Analyst. You will learn more about modeling with Spatial Analyst in Chapter 9 where you combine it with Model Builder. For now, you have a chance to see how to organize your data, set up your workspace, add raster data to your session, and save your maps.

Steps

Part I: Locating and Organizing Your Tutorial Data

1. Navigate to your installed data (C:\argcis\ArcTutor).

2. Right-click the Spatial Analyst folder and select Copy. The reason for this is so you don't corrupt the original dataset.

3. Browse to your working directory (e.g., C:\ drive, or E:\ drive). You might want to use a directory on an external hard drive so you can easily move from computer to computer with your data.

4. Right-click C:\ (or whatever workspace location you chose) and select Paste. This creates a complete copy of your database in that location. *Note:* You may have already discovered that it's a bit difficult to move from one place to another. That's because without copying and pasting, you are merely moving the software pointers, not the actual data.

Part II: Starting ArcMap

1. Start ArcMap by either using your Start menu in Windows or by double-clicking a shortcut on the system tray or your desktop screen.

2. If you have maps left open from your last session, shut them down.

3. Click on New Maps in the ArcMap—New Documents window if it isn't already highlighted.

4. Click the Open button.

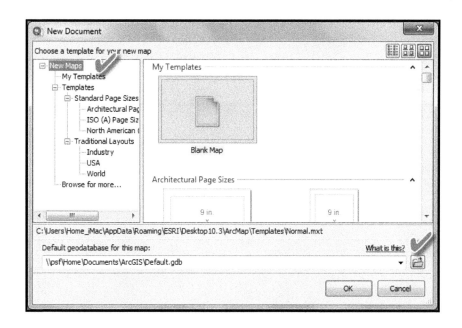

5. Click the Connect to folder button in the Select the map's geo-database window.

(continued)

ARCGIS LESSON 6-2 | (CONTINUED)

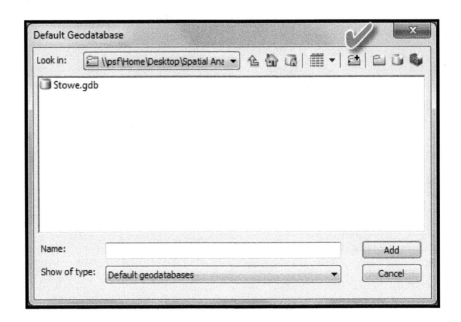

6. Navigate to the working copy of the Spatial Analyst folder you just created. *Note:* The figure above shows that I put mine on my desktop.

7. Click OK.

8. Click on the New File Geodatabase button. This will allow you to create a database with all the features you need to do your work.

9. Name the new file geodatabases Scratch.

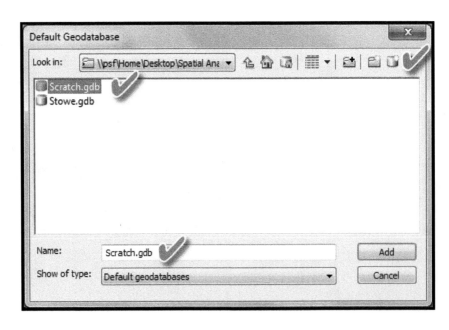

10. Click Add.

11. Click OK.

Part III: Setting up a Workspace

Your current and scratch workspaces are set to your Scratch.dbg geodatabase. What you are going to do is use the Stowe.gdb to access your data (this is the Spatial Analyst folder that contains your data), and then you will use the Scratch.gdb as your output folder.

1. Click the menu Geoprocessing > Environments.

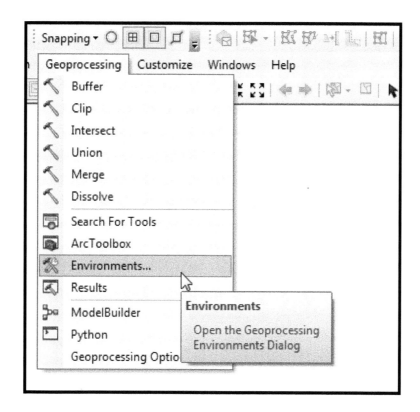

2. Click on Workspace to expand the environment settings related to your workspaces.

3. Under Current Workspace, navigate to the Stowe.gdb in the Spatial Analyst folder you created.

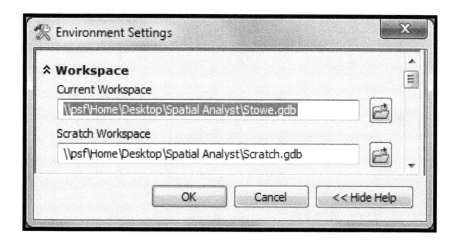

(continued)

ARCGIS LESSON 6-2 | (CONTINUED)

---

4. Click Add.

5. If not already set, navigate to your Scratch.gdb geodatabase.

6. Click OK.

Part IV: Adding Data to Your Session

1. At the top of the Catalog window, click the Toggle Contents Panel button 📇 until you see both the Catalog Tree and the Contents panel.

2. In the Catalog Tree, click on Stowe.gdb.

3. Notice how the ArcMap Table Of Contents now contains the selected databases (four feature classes and two rasters).

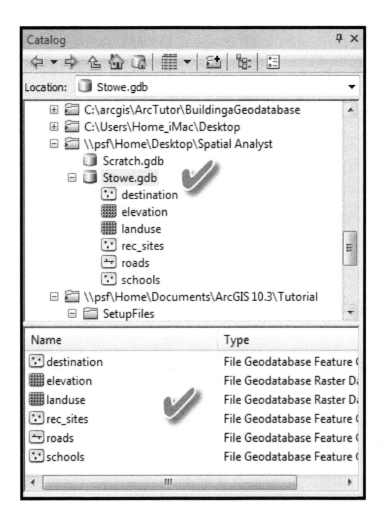

Part V: Saving Your Map

1. On the Standard toolbar, click the Save button 💾.

2. For the File name, enter Site Analysis.mxd.

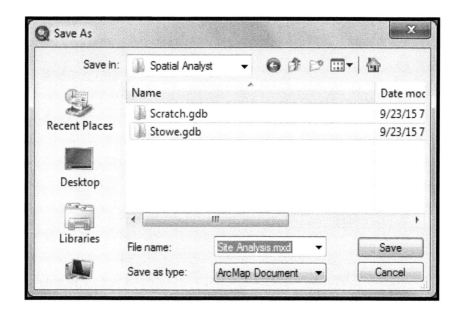

3. Save.

4. Take a screenshot of this to turn in to your instructor.

ADDITIONAL READING AND RESOURCES

DeMers, Michael N. *GIS Modeling in Raster*. New York: Wiley, 2001.

Tomlin, C. Dana. *GIS and Cartographic Modeling*. Redlands, CA: Esri Press, 2012.

KEY TERMS

GeoTIFF: A graphics (TIFF) file with embedded georeferenced information.

Image Analysis toolkit: A set of ArcGIS tools for manipulating and analyzing digital images such as satellite images and scanned aerial photography.

Map Algebra: A map-based language designed to allow the user to employ a wide range of mathematical, logical, and statistical techniques to raster GIS data.

Mosaic Dataset: A dataset that allows you to store, manage, view, and query various sizes of raster and image data. The dataset's function is to catalog images and rasters so they can be accessed easily and retrieved quickly.

syntax: The arrangement of words and phrases to create well-formed sentences in a language.

CHAPTER 7

Network Analysis

LEARNING OBJECTIVES

Here is the content you will learn in this chapter:

1. The types of analysis associated with network analysis.

2. The definitions of network dataset and geometric dataset, and identification of the components of these datasets.

3. Articulation of the layers that can be associated with network analysis for modeling.

4. How to use Network Analyst to create a network dataset and find the best route.

5. How to use Network Analyst to find the closest fire stations.

6. How to use Network Analyst to calculate service areas and create an origins–destinations (OD) cost matrix.

7. How to create a model for route analysis and service a set of orders with a fleet of vehicles.

8. How to use Network Analyst to perform location-allocation.

9. How to use Network Analyst to configure live traffic and perform network analysis using traffic data.

10. How to use Network Analyst with restriction attributes.

BEHAVIORAL INDICATORS

When you are finished with this chapter, you will be able to:

1. Provide a list of analyses associated with network analysis.

2. Define what a network dataset is in your own words.

3. Articulate the layers that can be associated with network analysis for modeling and what they mean.

4. Use Network Analyst to create a network dataset and find the best route.

5. Use Network Analyst to find the closest fire stations.

6. Use Network Analyst to calculate service areas and create an (origins–destinations) OD cost matrix.

7. Create a model for route analysis and servicing a set of orders with a fleet of vehicles.

8. Use Network Analyst to perform location-allocation.

9. Use Network Analyst to configure live traffic and perform network analysis using traffic data.

10. Use Network Analyst using restriction attributes.

Chapter Overview

This chapter introduces you to the basics of Network Analysis in GIS. Until now, you have been focusing on area-based thematic map layers. A substantial amount of data used in GIS, however, is based on linear objects such as rail lines, roads, pipelines, electrical lines, and even wildlife corridors. Because they are linear, they present different challenges and different opportunities for modeling. You will get a chance to learn about and work with such linear objects in this chapter.

A network is a collection of linear objects, nearly always composed of the same kinds of features. Networks come in two types, geometric and transportation. Geometric networks are those commonly associated with pipelines and electrical lines and are represented in GIS as a set of connected **edges** and **junctions** plus associated rules. Transportation networks are normally used to model transportation systems and are therefore typically associated with roads and railroads although some natural corridor studies have been performed using network datasets. In this chapter, you will focus on the traditional uses of both geometric and **transportation networks** and will get a chance to practice creating them for their intended use.

Geometric Networks

Geometric networks are collections of edges and junctions defined in an ArcGIS database together with rules for connectivity. The networks' purpose is to represent and model the behavior of nontransportation-related real-world infrastructure. In particular, the **geometric network** is the GIS representation of water distribution systems, electrical lines, telephone services, and stream flows. The network allows a person to be able to model and analyze the conditions and flows along the network. It includes the ability to trace streams downstream or upstream to determine the source of flow within a network, to measure demands for electricity, and to manage telephone and 9-1-1 services. Although there are some similarities between the transportation network and the geometric network, they are designed for different processes.

Geometric networks are composed of two elements, edges, and junctions. Edges are special cases of geodatabase line features inside feature datasets that have behaviors associated with utilities. They are designed to represent real-world water mains, electrical transmission lines, gas pipelines, and telephone lines. Edges come in two types, simple and complex. Simple edges are specifically designed to permit resources (water or electricity, for example) to enter one end of the edge and exit only at the other end of the edge. Complex edges allow resources to flow from one end to the other, like simple edges, and to be siphoned off along the edge without having to physically split the edge feature. This latter behavior allows for mid-span connectivity such as adding new junctions to be "snapped" into the span. Junctions are features that allow two or more edges to be connected and allow the transfer of flow and resources between edges. Examples of junction types include fuses, switches, service taps, and valves.

There are seven primary things for which the geometric network is designed to analyze. The first is to calculate the shortest path between any two points in the network. Utility companies use this approach to determine how to reroute water flows if there are blockages. Electrical companies employ the approach to redirect electricity if parts of the network go down.

To do what is considered in the last paragraph requires an understanding of how the network is connected. This is the second thing for which the geometric network is designed: to find all the connected and disconnected network elements. Utility companies use this approach to inspect the functionality of the system on a section-by-section basis to verify that each is operating at peak efficiency. This is as true of electrical transmission lines as it is of water lines. If there is a breach in a water line, the company can quickly identify exactly which section is the problem. A utility company can track a break in the transmission of electricity to the problem section.

A third, related functionality is the determination of flow direction of the edges when sources or sinks (where the water will collect) are set. Using this function, engineers will be able to identify the direction of flow along the edges, and ArcGIS can use the flow direction function to perform flow-specific network analysis. By this, I mean that selected portions of the system—let's say all the water flowing toward a certain reservoir—can be analyzed in case there is some pollutant being carried from the source to that location.

The fourth thing for which the geometric network is designed to analyze is very closely related to the determination of flow direction in that the network allows the user to trace the elements of the system both upstream and downstream to determine the availability of shutoff valves. The fifth capability is to calculate the shortest path both upstream and downstream from point to point to identify the source of pollution.

Electric utility companies will often use the locations of individual customer phones to determine possible transformer issues or downed power lines. This capability (number 6) is to be able to find all network elements upstream (up flow) from many points (in this case, the customer phones) and to determine (capability 7) which elements are common (e.g., individual power line segments or transformers).

ACTIVITY 7-1   GEOMETRIC NETWORKS

In this activity, you will get an opportunity to demonstrate your knowledge of networks, specifically geometric networks.

1. In your own words, explain what a network dataset is and what its primary uses are.

2. What is the difference between a geometric network and a transportation network?

3. In your own words, define edges and junctions. Explain how they differ from traditional line and point entities.

4. What is the difference between simple and complex edges, particularly with regard to behaviors?

5. Describe what the seven functions of a geometric network are and provide a quick example of each.

ARCGIS LESSON 7-1 | CREATING GEOMETRIC NETWORKS

This lesson gives you the opportunity to create a geometric network dataset. It will also allow you to examine firsthand junctions and simple and complex edges and how to create connectivity rules for use by these features.

Note: The first four parts of this exercise are needed to prepare your environment prior to building your geometric network.

Part I: Connecting to and Exploring Data

Through the ArcGIS Catalog, data are accessed through either folder or database connections. Database connections are used to access ArcSDE geodatabase.

The following exercise uses file geodatabases (a simple form of geodatabase) that are accessed through folder connections.

Other data you can access through folder connections include personal geodatabases, shapeless, and coverages.

When you examine a folder connection, you see folders and data sources immediately. This exercise will allow you to begin organizing your data by creating a folder connection to it via ArcCatalog.

Steps

1. Start ArcCatalog.

2. Click the Connect to Folder button ⬚ on the ArcCatalog Standard toolbar. This will open the Connect to Folder dialog box. Note that if you don't do this step, you will find the folders appearing empty because ArcGIS doesn't see the files without the connection.

3. Navigate to the BuildingaGeodatabase folder where your tutorial data are located.

4. Click **OK** on the Connect to Folder dialog box to establish a folder. Your new folder connection is now listed in the Catalog Tree. Now you can access all the data needed for this exercise. Before you start modifying the geodatabase, it is a good idea to explore it first to see how it's put together.

5. Click the plus sign next to the BuildingaGeodatabase folder to see the datasets.

6. Click the laterals to select it.

7. Click the Preview tab to view the laterals' geometry.

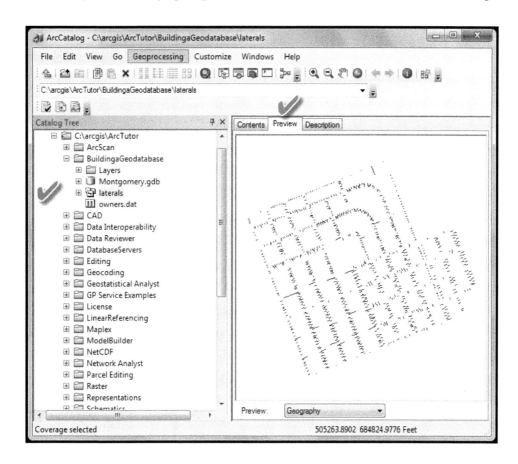

8. Click the plus sign next to the Montgomery geodatabase and then double-click each feature dataset. This expands the feature dataset so the feature classes are shown.

9. Click the owners.dat INFO table in the BuildingaGeodatabase folder. Notice that the Preview type automatically changes to Table (bottom of the table) and displays the table records. This table contains the owner information for the Parcels feature class in the Montgomery geodatabase. Next you will import the table in the geodatabase and create relationships between the parcels and their owners.

Save some screenshots for your instructor. Don't shut your system down yet.

Part II: Importing a Coverage

Steps

1. In ArcCatalog, right-click the Water feature Dataset in the Montgomery geodatabase, point to Import, and then click Feature Class (multiple).

This tool allows you to specify your input coverage, output geodatabase, and output feature class. Fortunately, because you opened this tool by right-clicking a feature dataset. The output geodatabase, Montgomery, and feature dataset, Water, are already filled in for you.

There are several ways to set the input and output datasets, and you can drag a dataset or datasets from the ArcCatalog

(continued)

ARCGIS LESSON 7-1 | (CONTINUED)

Tree or Contents tab and just drop them on the text box. You can also click the Browse button to open the ArcCatalog mini-browser and navigate to the desired dataset. Finally, you can type the full path name to the dataset in the text box.

2. Click the Browse button, navigate to the arc feature class in the laterals coverage, and click Add.

3. Click OK to run the Feature Class to Geodatabase (multiple) tool. A progress bar will appear in the lower right of Arc-Catalog. When the tool finishes the import, a pop-up message appears. You can click this message to open the Results pane (appears on the left) to see the generated messages. The laterals_arc feature class has now been added to the Water feature dataset.

4. In the ArcCatalog Tree, navigate to and click the laterals_arc feature class that you just added to the geodatabase.

5. Right-click and scroll to the rename function (or click F2) and rename the feature class to Laterals.

6. Click the Preview tab on the top to see the features.

Part III: Creating Aliases for Feature Classes and Fields

Aliases allow you to have multiple names for feature classes and fields that don't require you to follow the databases' object name limitations. Aliases also provide easy ways for you to remember what feature classes and fields become numerous. Fortunately, the geodatabase allows you to create these aliases.

When you use aliases in ArcMap, the alias name will automatically be used for feature classes, tables, and fields. However, in ArcCatalog, the feature classes, tables, and fields are always represented by their true names.

Now you will get a chance to create these aliases for the new feature class (Laterals) you just created.

Steps

1. Right-click the Laterals feature class in the Water feature dataset and click Properties.

2. Click the General tab.

3. In the Alias text box, type "Water laterals."

4. Click the Fields tab.

5. Click the OBJECTID field and then click on the bottom type Field identifier in place of the OBJECTID in the cell for its alias.

6. Click Apply.

7. Repeat the process, assigning the following aliases.

1. Field	2. Alias
3. Shape	4. Geometry field
5. DEPTH_BURI	6. Depth buried
7. RECORDED_L	8. Recorded length
9. FACILITY_I	10. Facility identifier
11. DATE_INSTA	12. Installation date
13. TYPECODE	14. Subtype code

8. Click OK when you have finished adding the aliases to close the Feature Class Properties dialog box.

Part IV: Importing the INFO Table

The owner.dat INFO table contains owner information for the parcels in the Parcels feature class in the Montgomery geodatabase. To be able to create relationships between the parcels and their owners, the owner information must be imported into the Montgomery geodatabase. To do this, you will be using the Table (single) import tool to import the owner.dat INFO table into the Montgomery geodatabase. You will then create aliases for the table.

Steps

1. Right-click the Montgomery geodatabase, point to Import, and then click Table (single).

2. Drag and drop the owners.dat INFO table from the Catalog Tree to the Input Rows text box of the Table to Table dialog box.

3. Type Owners in the Output Table text box.

4. Click OK.

5. When the tool is finished, click the Owners table in the Montgomery geodatabase in the Catalog Tree.

6. Click the Preview tab.

7. Right-click the Owners table and click Properties to see the properties of the table.

8. For the alias for this table, type "Parcel owners." Click Apply.

9. Click the Fields tab and type the following field aliases:

Field	Alias
OBJECTID	Object identifier
OWNER_NAME	Owner name
OWNER_PERCENT	Percentage ownership
DEED_DATE	Date of deed

10. Click OK.

The data in the laterals coverage and owners.dat INFO table are now in the Montgomery geodatabase for you to use. Now you will be able to leverage the geodatabase by assigning behaviors to your data. To do this, you need to create subtypes and attribute domains.

Part V: Creating Subtypes and Attribute Domains

One cool thing about storing your GIS data in a geodatabase is that you can decide on the rules about how the data can be edited. You define the rules by creating new attribute domains, creating subtypes for feature classes, and associating the new domain, existing domains, and default values and fields for each subtype. You will do that for the lateral diameters.

Attribute domains are the rules that define the acceptable values of a field type. Multiple feature classes and tables can share attribute domains stored in the database. However, not all the objects in a feature class or table need to share the same attribute domains.

For example, in a water network, suppose that only hydrant water laterals can have a pressure between 60 and 100 psi whereas service water laterals can have a pressure between 50 and 75 psi. You would use an attribute domain to enforce this restriction. To implement this kind of validation rule, you don't have to create separate feature classes for hydrant and service water laterals, but you would want to distinguish these types of water laterals from each other to establish a separate set of domains and default values. You can do this using subtypes.

For this exercise, you are going to use ArcCatalog to create a new coded value attribute domain. This new domain will describe a set of value pipe diameters for your newly created Laterals feature class.

Steps

1. Right-click the Montgomery geodatabase and click Properties. This opens the Database Properties dialog box.

2. Click the Domains tab.

3. Click the first empty field under Domain Name and type "LatDiameter" for the new domain name.

4. In the Description field, insert "Valid diameters for water laterals."

Here's your chance to specify the properties of the domain for LatDiameter. The properties you will specify include the type of field with which the domain can be associated, the type of domain it is (range or coded value), the split and merge policies, and the valid values you can include in the domain fields.

5. Under Domain Properties, click the drop-down menu for the Field Type and click Float (floating point numbers). This determines the data type of the column to which the domain can properly be applied.

6. Now click the drop-down menu for the Domain Type and click Coded Values.

7. Type the valid values, or codes for the coded value domain, and for each code, provide a user-friendly description. As you will see later in the tutorial, ArcMap uses the user-friendly description, not the code, for values of fields that have coded value domains associated with them.

8. Click the first empty field in the Code column under Coded Values: and type 13".

9. Click the Description field beside it and type 13" for the description.

10. Add the following coded values to the list:

Code	Description
10	10"
8	8"
6	6"
4	4"
3	3"
2.25	2 1/4"
2	2"
1.5	1 1/2"
1.25	1 1/4"
1	1"
0.75	3/4"
−9	Unknown

11. Click OK to close the dialog box. The domain has now been added to the geodatabase.

Part VI: Creating Subtypes and Associated Default Values and Domains

Your next step is to create subtypes for the Laterals feature class and associate default values and domains with the fields for each subtype. By creating subtypes, not all the water lateral features need to have the same domains, default values, or connectivity rules. This gives you some flexibility in setting up your database.

(continued)

ARCGIS LESSON 7-1 | (CONTINUED)

Steps

1. Double-click the Water feature dataset in the Montgomery geodatabase to open it.

2. Right-click the Laterals feature class and select Properties to open the Feature Class Properties dialog box.

3. Click the Subtypes tab.

   You are now going to specify the subtype field for this feature class. The subtype field contains the values that identify to which subtype a feature belongs.

4. Click the Subtypes Field drop-down arrow and click TYPECODE.

   You are now going to add subtype codes and descriptions. Note that when you add a new subtype, you assign default values and domains to some of its fields.

5. Click the Description field next to subtype code 0 under Subtypes and type "Unknown" for its description.

6. Under Default Values and Domains, click the Default Value field next to the H_CONFID and type "0" for its default value.

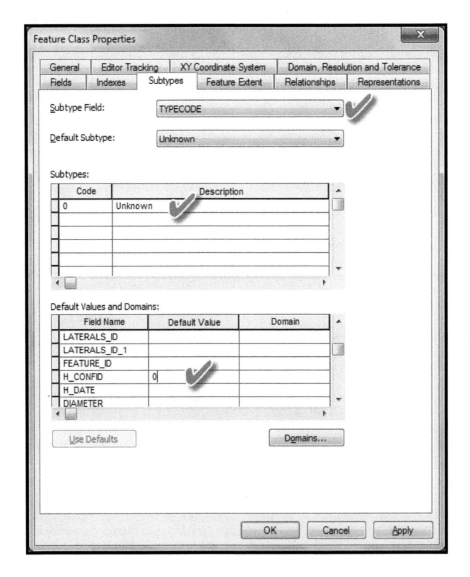

7. Type "0" for the default value of the DEPTH_BURI and RECORDED_L fields.

8. For the WNM_TYPE and PWTYPE fields, type "WUN-KNOWN" as the default values.

9. Click the Default Value field next to DIAMETER field and type "8" for the default value.

10. Click the Domain drop-down list for the DIAMETER field and click LatDiameter to set it as the default attribute domain for the Unknown subtype.

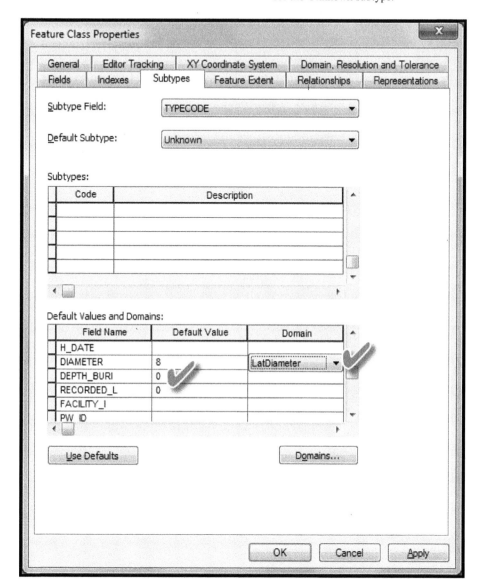

11. Click the Material field and type DI for the default value.

12. Click Material in the Domain drop-down list for Material.

13. Add the following subtypes:

Subtype code	Description
1	Hydrant laterals
2	Fire laterals
3	Service laterals

14. Set the default values and domains for the DEPTH_BURI, RECORDED_L, DIAMETER, and MATERIAL fields for each of these new subtypes the same as you did for the Unknown subtype.

15. For the Hydrant laterals subtype, set the WNM_TYPE and PWTYPE fields to WHYDLIN.

16. For the Fire laterals subtype, set the default values of the WNM_TYPE and PWTYPE fields to WFIRELIN.

17. For the Service laterals subtype, set the default values of the WNM_TYPE and PWTYPE fields to WSERVICE.

(continued)

ARCGIS LESSON 7-1 | (CONTINUED)

18. When adding new features to a feature class with subtypes in the ArcMap editing environment, if you don't specify a particular subtype, the new feature will be assigned the default subtype. Once you have added all the subtypes for this feature class, you can set the default subtype from those you entered.

19. Click the Default Subtype drop-down arrow and click Service laterals to set it as the default subtype.

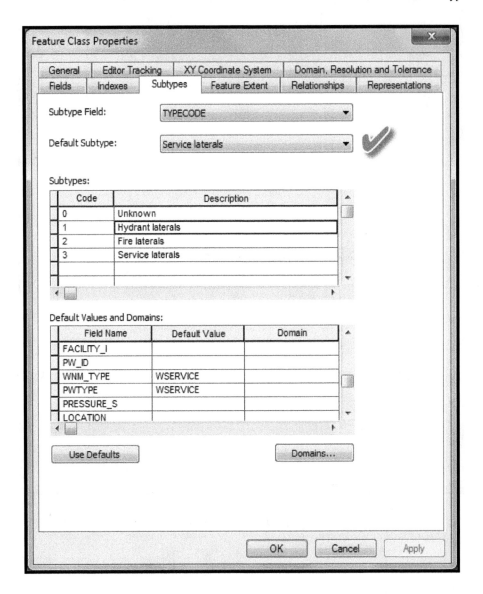

20. Click OK.

You have now added behaviors to the geodatabase by adding domains and creating subtypes. Next you will add some additional behavior to the geodatabase by creating relationships.

Part VII: Creating Relationships

Earlier you imported an INFO table with owner objects into the Montgomery geodatabase. The geodatabase already has a feature class, Parcels, that contains parcel objects. Now you're going to create a relationship class between the parcels and the owners so that when you use the data, you can determine who the parcel owners are.

Steps

1. Right-click the Landbase feature dataset in the Montgomery geodatabase, select New, and then click Relationship Class. The New Relationship Class wizard opens.

2. The first panel in the wizard allows you to specify the name, origin, and destination feature class or table for your new relationship class.

3. Type "ParcelOwners" in the Name of the relationship class text box.

4. Click Owners in the Origin table/feature class list.

5. Double-click the Landbase feature dataset in the Destination table/feature class list.

6. Click Parcels. This designates the Parcels feature class as the destination feature class.

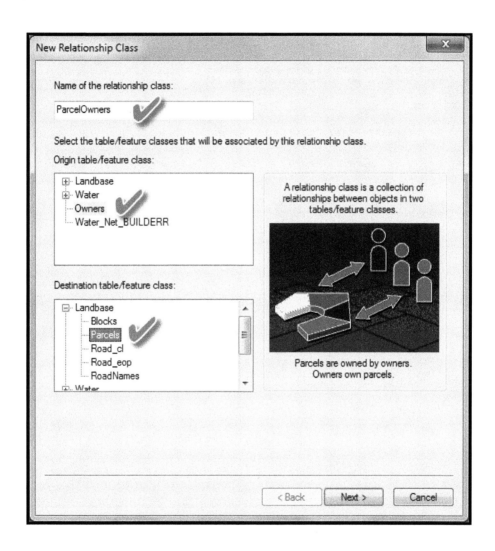

7. Click Next.

The next panel allows you to specify the type of relationship class you are going to create. For this lesson, you will create a simple relationship class because owners and parcels can exist in the database independently of each other. You can, therefore, accept the default type, Simple (peer-to-peer) relationship.

8. Click Next.

You now need to specify the path labels and the message notification direction. The forward path label describes the relationship as it is navigated from the origin class to the destination class—in this case, from Owners to Parcels. The backward path label describes the relationship when navigated in the other direction—from Parcels to Owners.

(continued)

ARCGIS LESSON 7-1 | (CONTINUED)

The message notification direction describes how messages are passed between related objects. Message notification is not required for this relationship class; therefore, you can accept the default of None.

9. Type "owns" for the forward path label.

10. Type "is owned by" for the backward path label.

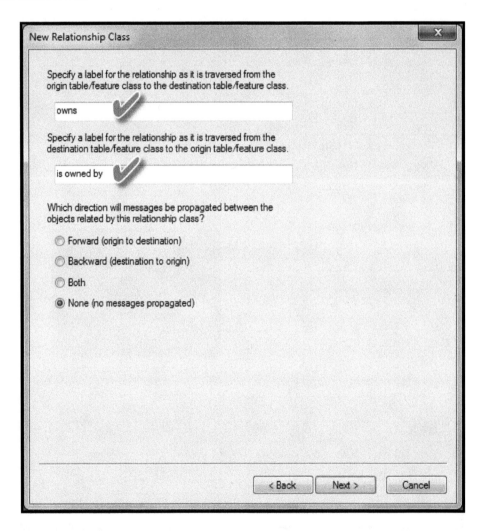

11. Click Next.

Now you are going to specify the cardinality of the relationship that describes the possible number of objects that can be related between the origin and destination feature classes or tables.

12. Click 1-M (one-to-many) to indicate that one owner can own many parcels.

13. Click Next. Now you need to specify whether your new relationship class will have attributes (not needed in this case, so you can accept the default).

14. Click Next.

The next step is to identify the primary key (the primary item you will search for) in the origin table (Owners) and the embedded foreign key (a secondary item you will search for) field in the destination class (Parcels). Owners and Parcels that have the same value in these fields will be related to each other.

15. Click the first drop-down arrow under Select the primary key field in the origin table/feature class and click PROPERTY_ID.

16. Click the second drop-down arrow on the dialog box and click PROPERTY_I for the embedded foreign key in the destination feature class.

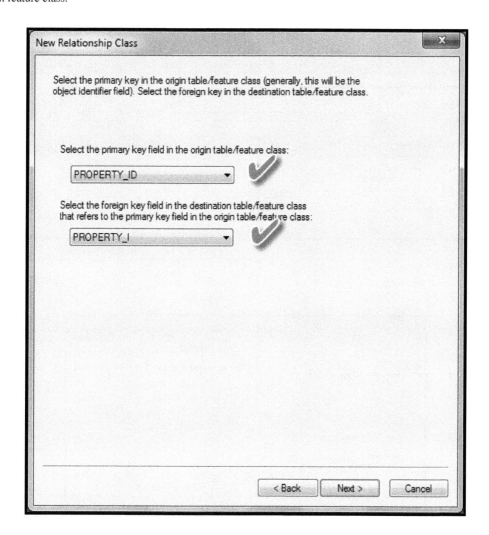

17. Click Next. A summary page appears.

18. Review the summary page to make sure the information is correct.

19. Take a screen shot for your instructor.

20. Click Finish.

21. Shut down ArcCatalog. This is a good place to take a break if you wish.

Part VIII: Creating a Geometric Network

Now all the pieces you need are in place for you to create your geometric network.

Steps

1. Start ArcCatalog.

2. Navigate to the Montgomery geodatabase in your Folder Connections. *Hint:* You will find it in the BuildingaGeodatabase folder in ArcTutor.

3. Expand the Montgomery geodatabase to expose the Landbase and Water features.

(continued)

ARCGIS LESSON 7-1 | (CONTINUED)

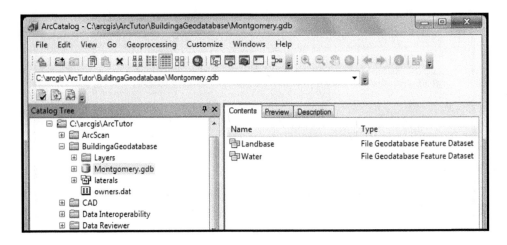

4. Right-click on the Water feature dataset and point to New, and then click Geometric Network.

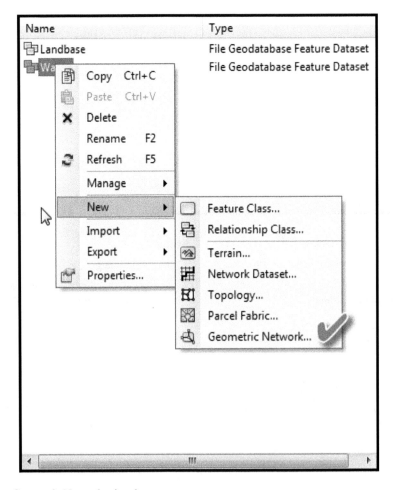

This opens the New Geometric Network wizard.

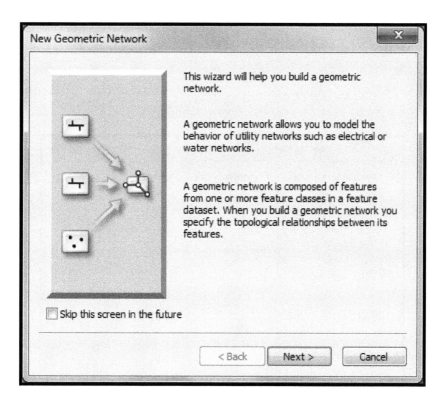

5. Click Next.

6. Type "Water_Net" for the name of the geometric network you are creating.

7. Select Yes to snap features.

8. Type "0.5" in the text box next to Feet.

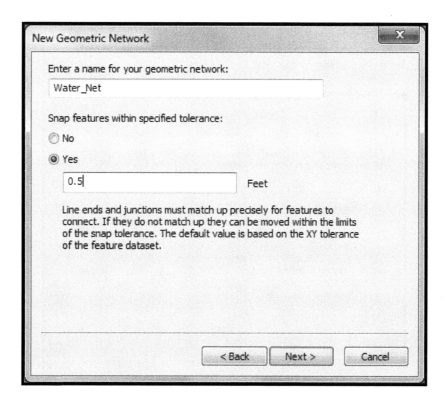

(*continued*)

ARCGIS LESSON 7-1 | (CONTINUED)

9. Click Next.

10. Now you can select the feature classes in the feature dataset that will be part of your geometric network.

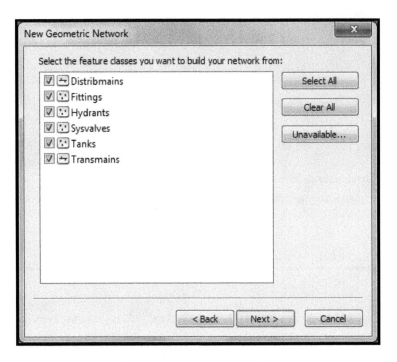

11. Click Next. You might want to exclude features that you don't need or don't want. This makes it easier to rebuild the network with the same elements if you want to start again.

12. Click No because you want to keep all the features of the network for this exercise.

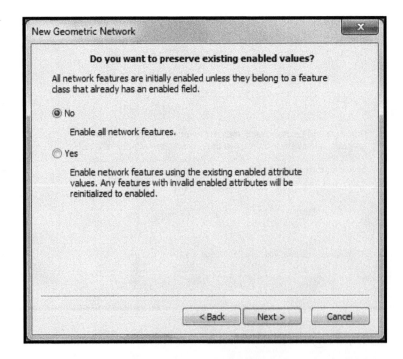

13. Click Next.

In the next dialog box, you get an opportunity to decide which line classes you are going to create as complex edge feature classes and which are going to remain simple edge feature classes. Recall from your reading what the difference is between these types. Remember also that the complex line remains a single edge but can have several access points. By default, line features are simple edge feature classes. Your second job is to decide which, if any, of the junction feature classes will be sources (start) and which will be sinks (end). This tells the network the flow direction (from source to sink).

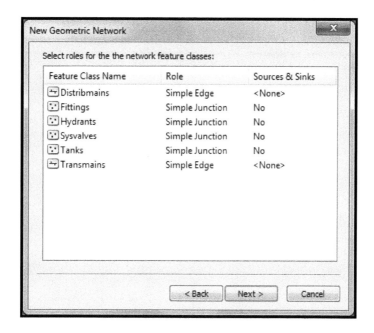

14. In the row for the Distribmains, select Simple Edge under the Role column.

15. Select Complex edge from the pulldown list. This changes the Distribmains feature class from the default (simple edge) to a complex edge.

16. Now in the row for the Tanks feature class, click the drop-down menu under Sources & Sinks, and select Yes.

17. In the row for Transmits, click Simple edge under the Role Column and then choose Complex edge from the list, changing their role from simple to complex edge.

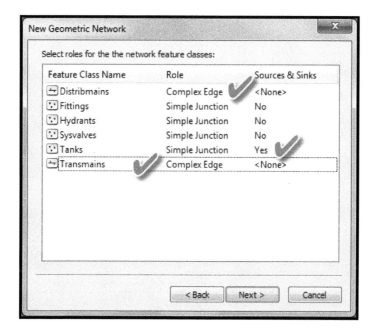

(*continued*)

ARCGIS LESSON 7-1 | (CONTINUED)

18. Click Next.

19. Now you are able to assign network weights (this describes the cost of traversing an element in the logical network such as a drop in water pressure as water flows through a pipe).

For this particular geometric network, weights aren't required (the default condition).

20. Click Next.

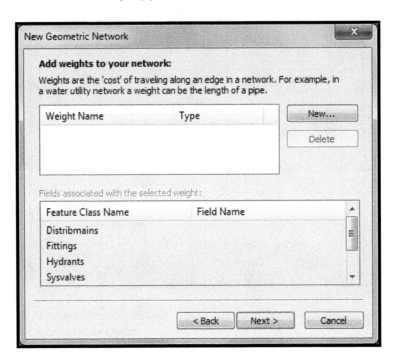

21. Click Next again to open the summary page.

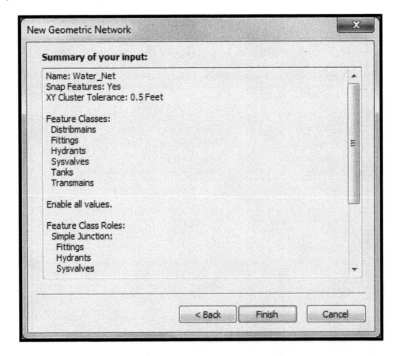

22. After you review the summary page, click Finish. Take a screenshot for your instructor.

    You will see a progress indicator for each stage of the network-building process.

    You will also receive an error message showing that the network was built but there were errors in the network. Don't panic!! These are left there on purpose.

23. Click OK to close the message box.

    You can see what errors occurred while building the geometric network by previewing the WaterNET_BUILDERR table.

24. Click the WaterNet_BUILDERR table in the Catalog Tree and click the Preview tab to view the entries in the table. Two records are displayed.

    Note: If you received more than two errors, delete the geometric network and repeat the steps to create it again. (Remember that all the steps in this exercise must be completed before creating the geometric network).

    Now you are going to add some connectivity rules. These constrain the type of network features that can be connected to each other and the number of features of any particular type that can be connected to features of another type. By establishing these rules, you can maintain the integrity of the network connectivity in the database. Take some screenshots along the way for your instructor.

25. Right-click the WaterNet geometry network in the Water feature dataset and click Properties. This opens the Network Properties dialog box that provides information about the feature classes in the network and a list of the network weights.

26. Click the Connectivity tab. This lets you add and modify connectivity rules to the geometric network.

    You will first create a new edge-junction rule that states that hydrants can connect to hydrant laterals; it also indicates that when a hydrant lateral is created, a hydrant junction feature should be placed at its free end.

27. Click the drop-down arrow and click laterals.

28. Click Hydrant laterals in the Subtypes in the feature class list.

    Now you will click the types of junctions that hydrant laterals are allowed to connect to in the network. For simplicity, hydrant laterals can only connect to hydrants.

29. Check Hydrants in the Subtypes of the network list. You should also specify that when you create a lateral, if an end of it isn't connected to another edge or junction, a hydrant is placed at the end.

30. Click the plus sign next to Hydrants in the Subtypes in the network list. The Hydrants subtype will expand.

31. Right-click Hydrants under the Hydrants subtype and click Set as Default. A blue D appears next to the hydrant subtype showing you that it is the default junction for this edge subtype.

    Now you'll create a new edge-edge rule that shows that hydrant laterals can connect to a distribution main through taps, tees, and saddles. The default junction for connections between hydrant laterals and distribution mains will be taps.

32. Click the plus sign next to Distribmains in the Subtypes in the network list to expand it.

33. Click Distribmains, which appears under the Distribmains subtype.

    Because you have checked an edge in the network subtypes list, the list of junction subtypes in the network becomes active. In this list, you can specify which junction types of hydrant laterals and distribution mains can connect through.

34. Click the plus sign to expand Fittings in the junctions subtype list.

35. Check Tap, Tee, and Saddle, in that order, under the Fittings junction subtype. Tap has a blue D next to it indicating that it is the default junction.

36. Still in the Junctions subtype list, check WaterNet_Junctions, which is the generic, or default network junction type.

37. Click OK. You're just created a functioning geometric network.

Transportation Networks

Transportation networks are actually not a separate network type, but I use the term here so you will understand that this operation is not about geometric networks. The correct term is actually the network dataset, and these datasets are specifically designed to model transportation networks in the real world. Network datasets are created from typical point and line features and include **turns** and connectivity. The toolset that you use for modeling with Network Datasets is the Network Analyst extension, and the analytics all happen on the network dataset.

The Network Dataset models the typical street networks and network conditions such as one-way streets, turn restrictions, speed limits, and overpasses/tunnels. The dataset is designed to allow for routing (e.g., the shortest path from place A to place B) and uses the rules set up based on the network conditions just mentioned. Another obvious feature of network datasets is the idea of **connectivity**, the idea that the lines on the dataset constitute a continuous feature that can be traversed throughout its length depending on the segment properties. In normal feature datasets, the lines are essentially unaware of each other except as entities; they have no understanding of the attribute values assigned to each line. Other characteristics that can be applied include **impedances** and hierarchy (e.g., road versus railroad at an intersection). The number of possibilities is quite large. To be able to model transportation, the Network Dataset keeps track of these conditions and a series of connectivity policies that include the relationships among junctions, line segments, and all the attribute values associated with each.

To construct a network dataset, they consist of three types of elements that create the network geometry and are responsible for the establishing. The elements are:

Edges. Like those in the geometric network from Activity 7.1, these are line features with additional attributes specifically designed for modeling these lines as transportation networks.

Junctions. Again, like junctions in the geometric networks in Activity 7.1, these junctions are point features that connect the edges and, in this case, include attributes to allow navigation from one edge to another.

Turns. Not included in the geometric networks, these store information and rules that affect the movement from edge to edge.

Collectively, these elements form the basic structure of the network. The network connectivity links the edges and junctions to each other. The turn elements are optional in that they are used when only they are needed; they store information about turning movement rules such as "no left turn" at particular junctions.

All of this activity requires the use of attributes in the network. These are the properties that control **traversability** over the network. Some examples include travel time of a given length of road, street restrictions for vehicles (e.g., hazardous cargo routes), speed limits, and one-way streets. There are five basic groups of properties for network attributes: name, usage type, units, data type, and use by default. The name attribute is self-evident (e.g., Street, Rail line), but the others require a bit of explanation.

- **Usage Type.** This determines how the attribute will be used during analysis and includes cost, descriptor, restriction, and hierarchy.

- **Units.** Usually for cost attributes, these include distance or time. The descriptors, hierarchy, and restrictions are generally without units because if there were any, they are unknown.

- **Data Types.** The available data types include Boolean, integer, float, or double (double precision). As you might guess, cost cannot be Boolean. Restrictions must be Boolean, and hierarchy is always integer.

- **Use by Default.** This automatically sets those attributes on a new network analysis layer. So, for example, if you set a particular cost as a default, whenever you create an edge that uses cost, the default you set will automatically be used. There is some restriction in setting defaults. *Only one cost attribute can be set in a single dataset as the default. Also, descriptor attributes cannot be used as defaults.*

The network attributes will be created when you start the New Network Dataset wizard or the Network Dataset Properties dialog box on the Attributes tab. To create the network attributes, you will first define the name of the attribute (e.g., Street) and its usage, units, and data type. Then you assign the evaluators for each source that you will use to provide the values for the network attributes when the network is built. To do this, you will select the attributes you are working with and select Evaluators.

Some of the terms used so far might be a bit vague, so I will give more detail here. The term *cost* applies to certain attributes used to model impedances (things that slow you down) such as travel time or demand (number of students a school bus can hold). These attributes can be apportioned along edges such that they can be accumulated along the distance of the edges rather than as a single value per edge segment. These values are used for actions like finding the fastest or shortest route or routing busses to pick up a certain number of students.

Descriptors are attributes that describe the characteristics of a network or its elements. Descriptors, unlike cost attributes, are not apportioned along the length of an edge. An example includes the number of lanes or type of road. While these can't be apportioned and can't be used directly as impedance values, they can be used in conjunction with distance to create cost attributes. An example is the type of highway that determines its speed limit.

Restrictions are used during analysis to limit the use of edges, especially limiting traversability. The classic case of this is the one-way street restriction that allows proceeding only in a particular direction. This will impact the results of shortest path analysis. Restrictions are Boolean in that the restriction either exists (you cannot travel a particular direction) or does not exist.

Hierarchy is the order or importance to which you assign network elements. You are familiar with different highway types (e.g., local roads, streets, U.S. highways, interstate highways). During the process of finding shortest path, you might purposely want to avoid interstate highways). *Note:* ArcGIS Network Analyst limits your class hierarchy to three road ranges (e.g., primary, secondary, and local). To get around this if you have more than three classes, you can reclassify them into your ranges when you create the network dataset. You will learn more about all of these when you perform the ArcGIS Lessons.

ACTIVITY 7-2   TRANSPORTATION NETWORKS

In this activity, you will get a chance to review what you learned about the Network Dataset (transportation networks).

1. What is a network dataset, and how does it differ from geometric networks?

2. What are the three basic elements of the network dataset, and what do they do?

3. Why is connectivity necessary for network datasets?

4. What are the five basic types of attributes used in a network dataset? In your own words, describe what they are used for.

5. In your own words, describe why restrictions are Boolean.

6. What is the purpose of apportioning cost values? Why can't they just be assigned uniformly to an edge?

ARCGIS LESSON 7-2 | WORKING WITH TRANSPORTATION NETWORKS

This lesson gives you the opportunity to create a transportation network dataset. It will also allow you to examine firsthand the features, rules, and behaviors that make transportation networks so powerful for modeling traffic flows.

Part I: Creating a Network Dataset

Steps

1. Start ArcCatalog.

2. Enable the Network Analyst extension (this is needed for transportation nets).

   Click Customize > Extensions. The Extensions dialog box will open.
   Check Network Analyst.
   Click Close to shut the dialog box.

3. On the Standard toolbar, click the Connect To Folder ⬛ button to open the Connect To Folder dialog box.

(continued)

ARCGIS LESSON 7-2 | (CONTINUED)

4. Navigate to the tutorial folder with the ArcGIS Network Analyst tutorial data. The default is C:\arcgis\ArcTutor\ Network Analyst\Tutorial.

5. Click OK. A shortcut to the folder is added to the Catalog Tree under Folder Connections.

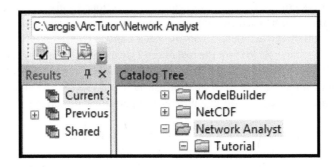

6. In the Catalog Tree, expand . . .\ArcTutor\Network Analyst\ Tutorial > Exercise01 > SanFrancisco.gdb

7. Expand SanFrancisco.gdb

8. Click the Transportation feature dataset. The feature classes

and feature dataset contents are listed on the Contents tab of ArcCatalog.

9. Right-click the Transportation feature dataset and click New > Network Dataset.

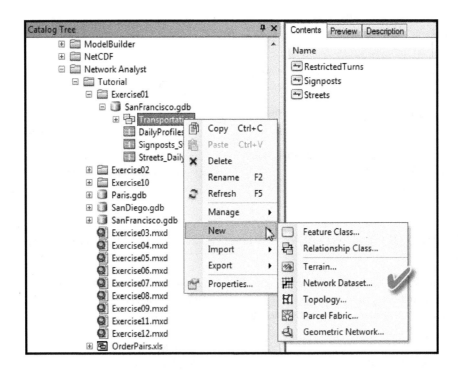

10. This opens the New Network Dataset wizard.

11. In the wizard, type "Streets_ND" for the name of the network dataset.

12. You will have an opportunity to choose the version of the network dataset. Use the default (10.1).

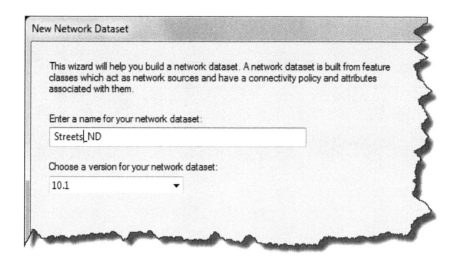

13. Click Next.

14. If it is not open, check the Streets feature class to use it as a source for the network dataset.

15. Click Next.

16. Click Yes to model turns in the network.

17. Check <Global Turns>, which enables you to add default turn penalties, and check RestrictedTurns to select it as a turn feature source.

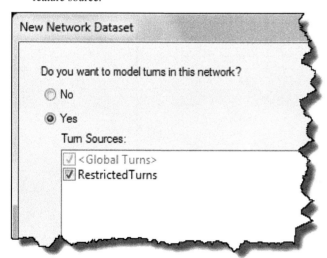

18. Click Next

19. Click Connectivity. The Connectivity dialog box opens. Here you can set up the connectivity model for the network.

20. For the Streets feature class, all streets connect to each other at endpoints.

21. Make sure that the connectivity policy of Streets is set to End Point.

22. Click OK to return to the New Network Dataset wizard.

23. Click Next.

24. Make sure the Using Elevation Fields option is selected because the transportation dataset includes elevation fields. This is critical because the transportation network between two places with different elevations must take that into account.

There are two ways to model elevation in a transportation network: using true elevation values from geometry or using logical elevation values from elevation fields.

The Streets feature class has logical elevation values stored as integers in the F_ELEV and T_ELEV fields. If two coincident endpoints have field elevation values of 1, for example, the edges will connect. However, if one endpoint has a value of 1 and the other coincident endpoint has a value of 0 (zero), the edges won't connect. ArcGIS Network Analyst recognizes the field names in this dataset and automatically maps them, as shown in the following graphic. (Only integer fields can serve as elevation fields.)

(continued)

ARCGIS LESSON 7-2 | (CONTINUED)

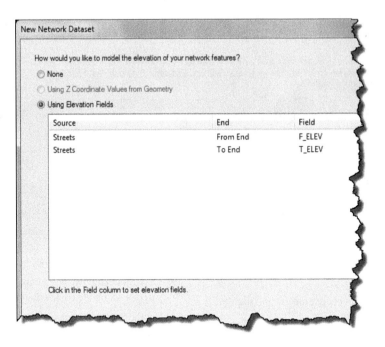

25. Click Next.

This page of the wizard allows you to enable things like quickest path based on time of day of week. Think, for example of the difference in travel time during rush hour and during normal times.

The San Francisco geodatabase contains two tables, DailyProfiles and Streets_DailyProfiles that store past traffic data. The schema of the tables was designed so the Network Analyst could recognize the role of each table and configure the past traffic automatically.

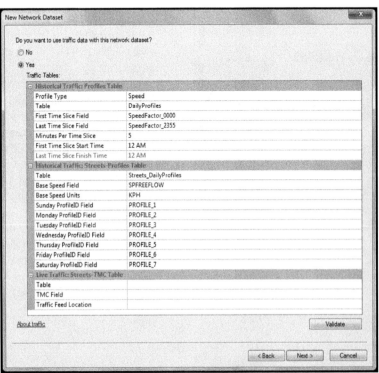

26. Click Next. The page for setting network attributes
    is displayed.

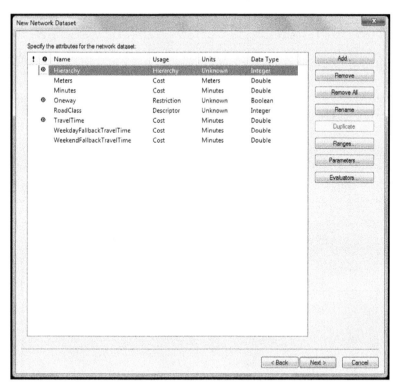

27. The attributes for the network are properties that control navigation including such aspects as impedances and restrictions that prohibit movement in one or two directions (depending on the type of road).

28. Click Evaluators.

    The reason for this is that Network Analyst identifies common fields such as Meters, Minutes (From-To, To-From), and others, and automatically populates them. In the case of the San Francisco data, the following eight attributes are set: (1) Hierarchy, (2) Meters, (3) Minutes, (4) Oneway, (5) RoadClass, (6) TravelTime, (7) WeekdayFallbackTravelTime, and (8) WeekendFallbackTravelTime. It also assigns evaluators to the attributes.

29. Click the Meters row to select it, and then click Evaluators to examine how the values of network attributes are determined. This opens the Evaluators dialog box.

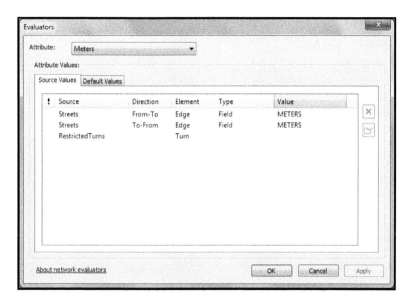

(*continued*)

ARCGIS LESSON 7-2 | (CONTINUED)

30. Notice the Source Values table to see the source feature classes. The source features (Streets) are listed twice (once for each direction defined by the direction of digitizing). The Element column defines the nature of the elements, in this case Edge for streets and Turn for RestrictedTurns. The Type column indicates the type of evaluator used to determine the network attribute values. Finally, the Value column shows information used by the evaluator to calculate attribute values.

31. From the Attribute drop-down list, click each attribute type, one at a time, and inspect the evaluator types and values for the source feature classes.

32. Now click OK to return to the NewNetwork Dataset wizard.

33. In the next few steps, you will add a new attribute to restrict movement over the turn elements created from the RestrictedTurns feature class.

34. Click Add to open the Add New Attribute dialog box.

35. Type "RestrictedTurns" in the Name field.

36. For Usage Type, select Restriction.

37. Note that Use by Default is checked. This will be applied as the default when you create a new network analysis layer. You can disable the restriction status if you want to.

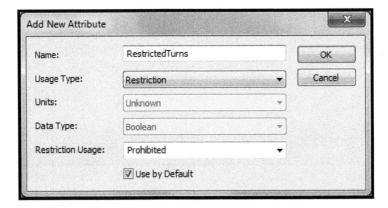

38. Click OK to add the new attribute, RestrictedTurns, to the list of attributes. The blue circle with the D indicates the attribute is enabled by default in the new analysis.

39. Click Evaluators to assign values by source to the new attribute.

   Perform the following steps to set the type of evaluator for RestrictedTurns to Constant with a value of Restricted.

40. Click the Attribute list and choose RestrictedTurns.

41. For the RestrictedTurns row, click under the Type column, and then choose Constant from the drop-down list.

42. Click the Value column and select Use Restriction.

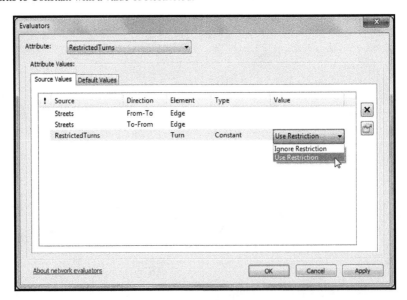

The evaluators for the street sources are empty, so they will remain in the Ignore Restriction condition when this restriction is used.

43. Click OK to return to the New Network Dataset wizard.

44. Now right-click the Hierarchy row and choose Use by Default. The blue symbol on the left disappears from the attribute, meaning that hierarchy will *not* be used by default when an analysis layer is created.

45. Click Next. If another menu requesting travel mode appears, ignore it and accept the defaults by hitting Next again to get to the directions setup menu.

46. Click Yes to set up directions.

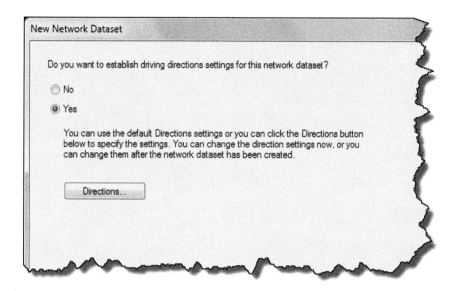

47. On the same menu, select Directions to open Network Directions Properties.

    Now you can specify the fields that will be used to report directions for any network analysis results.

48. On the General tab, be sure that the Name field for the Primary row automatically mapped to NAME. This field contains the street names for San Francisco that you need to generate driving directions.

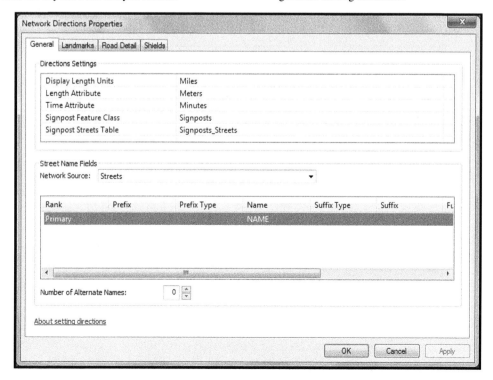

(*continued*)

ARCGIS LESSON 7-2 │ (CONTINUED)

49. Click OK and return to the New Network Dataset wizard.

50. Click Next. Ignore the Build Service Area Index menu.

51. Click Next again and the summary of your settings will display.

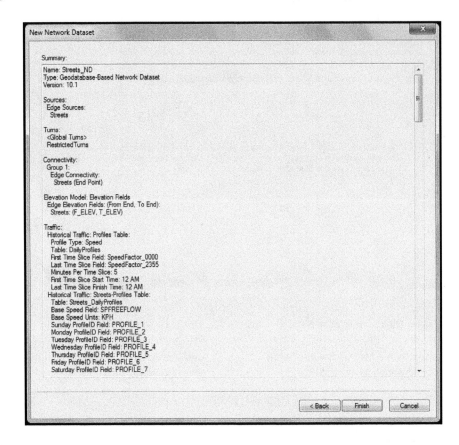

52. Click Finish.

A progress bar will open indicating the Network Analyst is creating the network dataset you configured.

Next the system will ask you if you want to build the network. This is necessary before any analysis can take place.

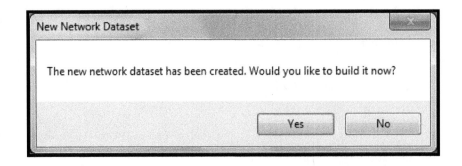

53. Click Yes.

The new dataset, Streets_ND, has been added to ArcCatalog together with the system junctions feature class, Streets_ND_junctions.

54. Click the dataset name and select the Preview tab to examine it. Take a screenshot for your instructor.

55. Close ArcCatalog. Preview the dataset.

ARCGIS LESSON 7-3 | FINDING CLOSEST FACILITY

This short lesson gives you the opportunity to continue modifying the network dataset you created and to do an analysis to find the closest facility.

Part I: Setting up Network Analyst
Steps

1. Start ArcMap.
2. Browse to your Network Analyst Tutorial Data.

3. Double-click Exercis04.mxd. This will open the map document.
4. Enable Network Analyst if it is not enabled (Customize > Extensions. Check Network Analyst. Close).
5. If the Network Analyst toolbar isn't displayed, add it.
6. To add the toolbar, click Customize > Toolbars > Network Analyst. The toolbar will be added to ArcMap.

7. If the Network Analyst window is not displayed, add it.
8. To add the window, on the Network Analyst toolbar, click the Show/Hide Network Analyst Window

button . The dockable Network Analyst window will open.

9. The Network Analyst window can be docked wherever you wish.

Part II: Creating the Closest Facility Analyst Layer
Steps

1. Click Network Analyst on the Network Analyst toolbar and select New Closest Facility.

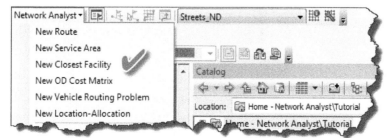

(*continued*)

ARCGIS LESSON 7-3 │ (CONTINUED)

This adds the closest facility layer to the Network Analyst window. Note that the Facilities, Incidents, Routes, Point Barriers, Line Barriers, and Polygon Barriers are all empty (0).

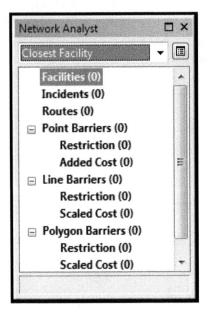

The analysis layer is also added to the Table Of Contents window.

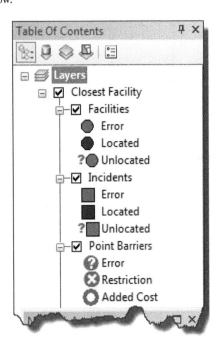

Part III: Adding Facilities

You will now load facilities from an existing point layer that represents fire stations.

Steps

1. In the Network Analyst window, right-click Facilities (0), and then click Load Locations. This opens the Load Locations dialog box.

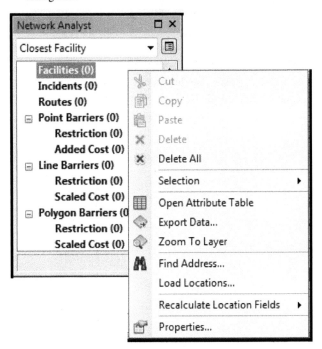

2. Choose FireStations from the Load From drop-down list.

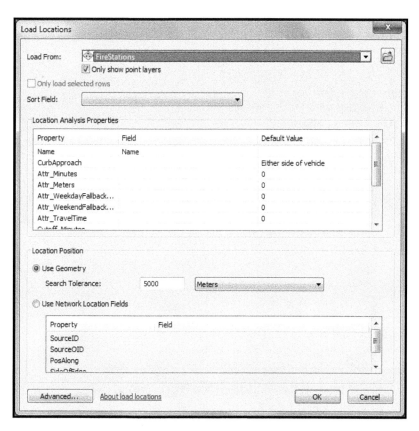

3. Click OK. Forty-three fire stations are displayed in the map as facilities and are listed in the Network Analyst window.

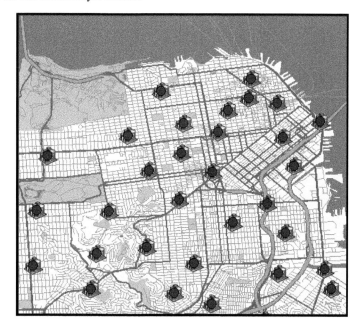

Part IV: Adding an Incident

Now you are going to add an incident by geocoding an address that comes in from an emergency call.

(continued)

ARCGIS LESSON 7-3 | (CONTINUED)

Steps

1. In the Network Analyst window, right-click Incidents (0) and choose Find Address. The Find dialog box opens.

2. Make sure you select SanFranciscoLocator in the Find Address. The Find dialog box opens.

3. In the Street or Intersection text box, enter 1202 Twin Peaks Blvd.

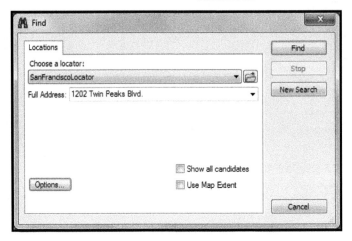

4. Click Find. The location with the Twin Peaks address is listed as a row in the table at the bottom of the Find dialog box.

5. Right-click that row and select Add as Network Analysis Object.

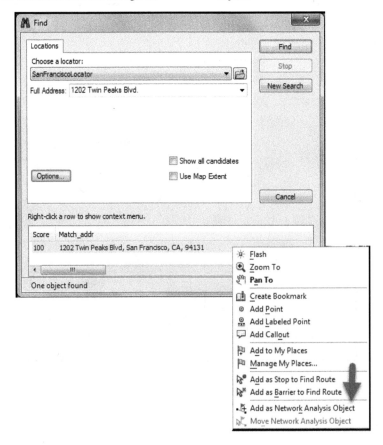

This adds the location address as an incident that is visible on the map and in the Network Analyst window.

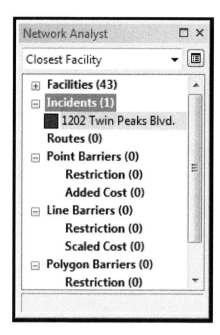

6. Close the Find dialog box.

Part V: Setting up Parameters for the Analysis

Now you can specify the parameters to analyze for the closest facility.

Steps

1. Click the Analysis Layer Properties button on the Network Analyst window. This opens the Layer Properties dialog box.

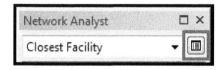

2. Click the Analysis Settings tab.

3. Make sure that the Impedance is set to TravelTime (Minutes).

4. Enter 3 in the Default Cutoff Value text box. ArcGIS will search for fire stations that are within 3 minutes of the Twin Peaks Boulevard fire location. Fire stations beyond the cutoff time are ignored.

5. Now increase Facilities to Find to 4. Why are you changing this value? The software will now search for a maximum of four fire stations from the site of the fire. The 3-minute cutoff still applies. If only three fire stations are within the 3-minute cutoff, then the program will search for a fourth fire station.

6. Choose Facility to Incident for the Travel From direction. The search results spread out from the fire stations that are located as facilities. This models the fire engines traveling from the stations to the fire (incident).

7. Click the U-Turns at Junctions drop-down arrow and choose Allowed.

8. Click the Output Shape Type drop-down arrow and choose True Shape with Measures.

9. Uncheck Use Hierarchy.

10. Check Ignore Invalid Locations.

11. In the Restrictions frame, uncheck RestrictedTurns. Fire engines don't have to obey traffic laws in emergencies.

12. In the Directions frame, make sure that Distance Units is set to Miles, Use Time Attribute is checked, and the time attribute is set to TravelTime (Minutes).

(continued)

ARCGIS LESSON 7-3 | (CONTINUED)

13. Your Analysis Settings should look like this:

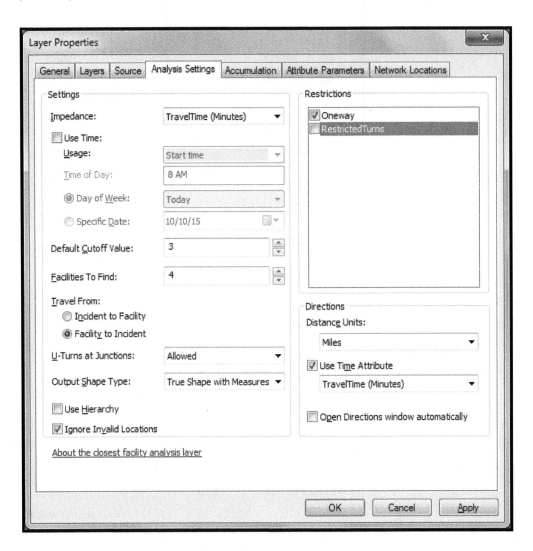

14. Click OK.

Part VI: Identifying the Closest Facilities

Steps

1. Click the Solve button ⊞ on the Network Analyst toolbar. Routes appear in the map display and the Route class in the Network Analyst window. Make a screenshot for your instructor.

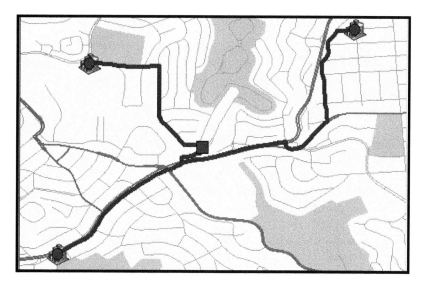

Note: Remember that you set your analysis up to find four facilities within the 3-minute distance; however, only three facilities are within that limit.

2. Click the Directions window button on the Network Analyst toolbar. This opens the Directions (Closest Facility) dialog box showing driving directions to the fire from each fire station. Make a screenshot for your instructor.

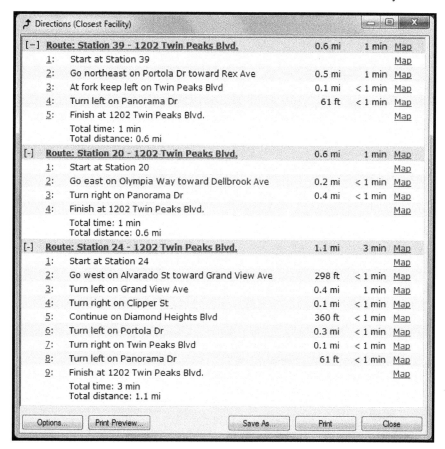

3. Shut down ArcMap.

4. When the menu pops up asking whether to save changes, click Yes.

Location Allocation

You might find it interesting that **location** and **allocation** are listed in the title of this section. While individually they are different, they are closely related. Location is a necessary prerequisite for allocation. Put another way location-allocation is a binary (twofold) process that simultaneously locates one or more facilities and then allocates the demand points to the facilities.

Location, the prerequisite process to location-allocation, is much like what the word implies in that you want to put a business, for example, in the best location. By *best location* here, this is the one that is most likely to bring you the most customers to generate the most income from your sales. It is also used to keep costs down. For example, facilities should be located in such a way that the costs of travel (usually measured in distance) to it are reduced as much as possible. Such facilities might be schools, fires stations, libraries, and emergency facilities.

There are six different problem scenarios for location-allocation: (1) minimize impedance, (2) maximize coverage, (3) minimize facilities, (4) maximize attendance, (5) maximize market share, and (6) target market share. The following ArcGIS lesson will give you a feel for a few of these scenarios and how to solve them.

ARCGIS LESSON 7-4 | WORKING WITH LOCATION AND ALLOCATION

This short lesson gives you the opportunity to perform a location-allocation analysis.

Part I: Starting Network Analyst

Steps

1. Start ArcMap.

2. Navigate to the Network Analyst\Tutorial data. The default location is C:\ArcGIS\ArcTutor\Network Analyst\Tutorial.

3. Double-click Exercise09.mxd. A map document will open in ArcMap.

4. If the Network Analyst extension is enabled, skip to step 8.

5. Click Customize > Extensions. The Extensions dialog box opens.

6. Check Network Analyst.

7. Click Close.

8. If the Network Analysis is displayed, skip this step as the toolbar will already be available.

9. To add the toolbar, click Customize > Toolbars > Network Analyst. The toolbar will be added to ArcMap.

10. If the Network Analyst window is not displayed, add it.

11. To add the window, on the Network Analyst toolbar, click the Show/Hide Network Analyst Window button. The dockable Network Analyst window will open.

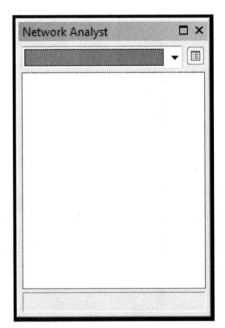

12. The Network Analyst window can be docked wherever you wish.

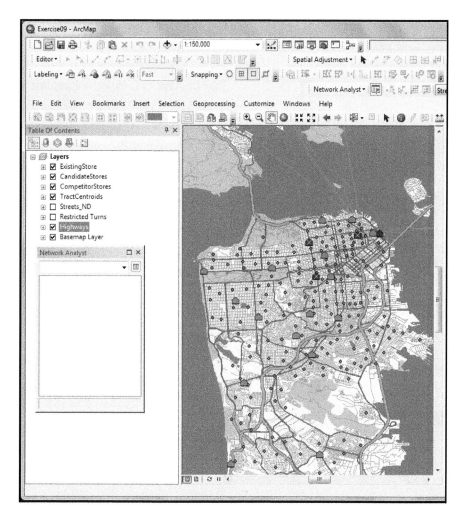

Part II: Creating the Location-Allocation Analysis Layer

Steps

1. Click Network Analyst on the Network Analyst toolbar and
   click New Location-Allocation.

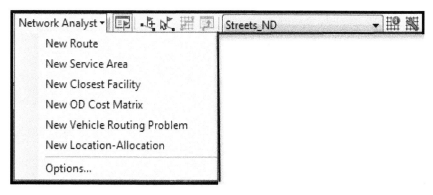

A location-allocation analysis layer is now added to the Network Analyst window with all network analysis classes (Facilities, Demand Points, Lines, Point Barriers, Line Barriers, and Polygon Barriers) being empty.

(*continued*)

ARCGIS LESSON 7-4 | (CONTINUED)

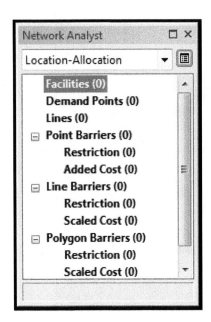

And the analysis layer is also added to the Table Of Contents.

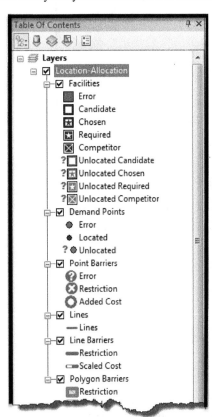

Part III: Adding Candidate Facilities

At this point, you would normally add candidate store locations (potential locations) to the network analysis class Facilities. These have already been added to the layer (CandidateStores) in the map. The candidate store names are in the layer's attribute table. Now you're going to load the point features from CandidateStores into the Facilities class of the location-allocation layer.

Steps

1. In the Network Analyst window, right-click Facilities (0) and choose Load Locations. This will load the Load Locations.

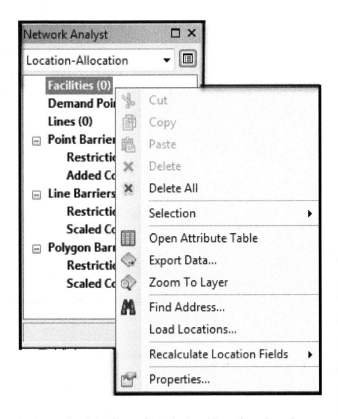

2. Select CandidateStores from the Load Locations drop-down menu to open the dialog box.

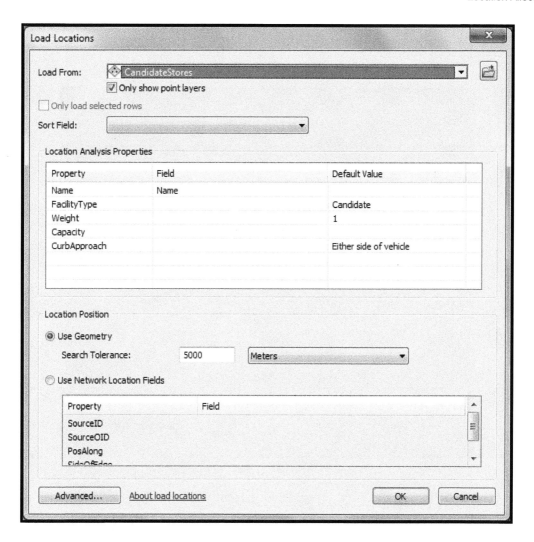

3. The Location Analysis Properties section of the Load Locations dialog box allows you to decide which attributes of the CandidateStores feature class contain the values that the Network Analyst will use to solve the location-allocation problem.

4. In the Location Analysis Properties section, make sure the Name property is automatically mapped to the Name field.

5. The Network Analyst attempts to match the location analysis property automatically for a newly created location-allocation layer based on a configure file (typically located in C:\Program Files\ArcGIS\Desktop10.3\NetworkAnalyst\NetworkConfigurations\NASolverConfiguration.xml).

6. Now click OK and the sixteen candidate stores will load into the Facilities network analysis class. The new facilities will be listed in the Network Analyst window and displayed on the map. Take a screenshot for your instructor.

(continued)

ARCGIS LESSON 7-4 | (CONTINUED)

Part IV: Adding Demand Points

The purpose of this analysis is to locate stores to best serve the existing population. For this reason, a point layer of census tract centroids has already been added to ArcMap. You will now load these centroids into the demand points network analysis class.

Steps

1. Now in the Network Analyst window, right-click Demand Points (0) and choose Load Locations.

2. Select TractCentroids from the Load Locations list.

3. In the Location Analysis Properties section, make sure the Name property is automatically mapped to the NAME field.

4. Click the Field column for the Weight property and choose POP2000. Each demand point will be weighted by the population as enumerated by the 2000 census.

5. Click OK. The 208 census tract centroids will load into the Demand Points class and are listed in the Network Analyst window and displayed on the map. Take a screenshot for your instructor.

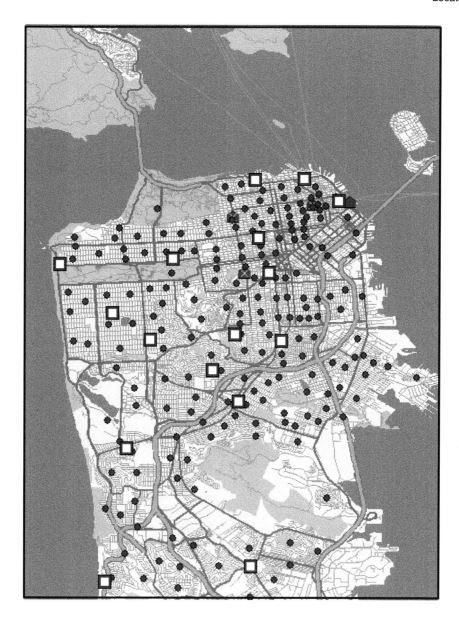

Part V: Setting up the Properties of the Location-Allocation Analysis

Steps

1. Click on the Analysis Layer Properties button on the Network Analyst window to open the Layer Properties dialog box.

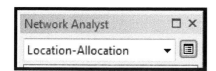

2. Click the Analysis Settings tab.

3. Make sure Impedance is set to TravelTime (Minutes).

4. Now set the Travel From to Demand to Facility. The default option would be a good choice for the minimize impedance and maximize coverage problem types. In the current situation, you want to maximize attendance and market share, and to target that market share. In such cases, demand tends to travel to the facilities, thus Demand to Facility is a better choice.

5. Click the U-Turns at Junctions drop-down arrow and select Allowed.

6. Set Output Shape Type to Straight Line. While the output will be displayed with straight lines, the travel costs will still be measured along the network.

(continued)

ARCGIS LESSON 7-4 | (CONTINUED)

7. Be sure that the Use Hierarchy and Ignore Invalid Locations are both checked.

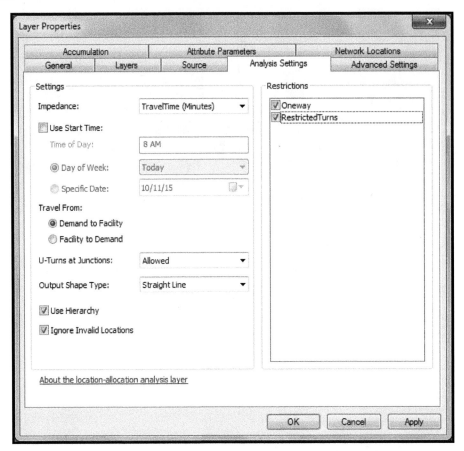

8. Now click the Advanced Settings tab.
9. From there, click the Problem Type drop-down list and select Maximize Attendance. You have just created a model based on the idea of maximizing attendance. It is a model because it assumes that all stores will have equal attraction and that people will choose the stores based entirely on nearness.

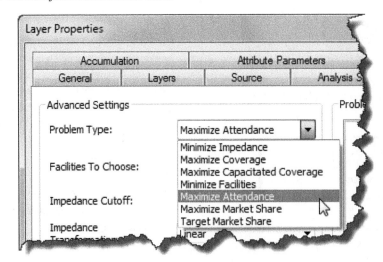

10. Now increase the Facilities to Choose to 3. The software will attempt to choose 3 facilities out of the 16 to serve the 208 demand points to the maximum possible.

Because you assume people won't travel more than 5 minutes to shop at the selected stores, you need to increase the value of Impedance to 5 because TravelTime is set to minutes.

11. Increase Impedance Cutoff to 5.

Be sure to set Impedance Transformation to Linear. This is the calculation of people's likelihood to visit a store. With a 5-minute impedance cutoff and a linear impedance transformation, the probability of a patron visiting a store decays at 1/5 of the total per minute. So, a store that is 2 minutes away from a demand point has a 60% probability of being visited.

One at 3 minutes away would have a probability of 40%, and so on. Obviously, this is an assumption, which is why this is called a model.

12. Click OK.

Part VI: Run the Process to Determine the Best Store Locations

Steps

1. Click the Solve button on the Network Analyst toolbar.

The lines on the map connect selected stores to their associated demand point locations. These lines also appear in the Lines class in the Network Analyst window. Take a screenshot for your instructor.

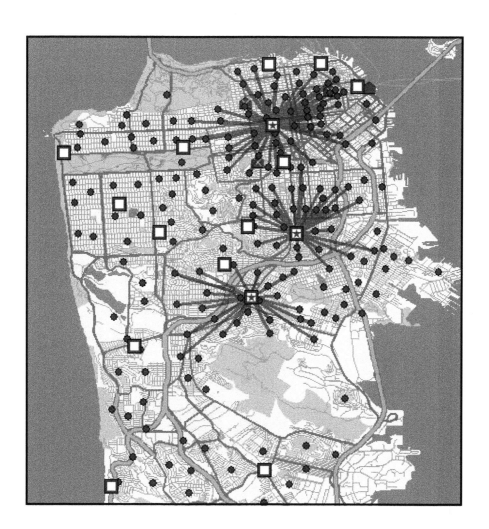

(*continued*)

ARCGIS LESSON 7-4 | (CONTINUED)

Now you will inspect the results in a bit more detail.

2. In the Table Of Contents, right-click the Facilities sublayer and select Open Attribute Table.

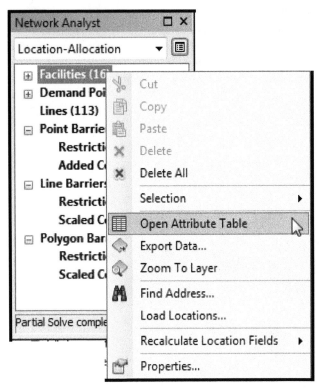

Notice in the Attribute Table that three features have their FacilityType field values set to Chosen rather than the default, Candidate. The DemandCount column lists the number of demand points for each of the three chosen facilities. If you add these up (25 plus 31 plus 57), you only have 113 demand points allocated to the chosen facilities. This is because some of the points were beyond the 5-minute maximum you set. The DemandWeight column lists the amount of demand allocated to each facility. The number indicates the number of people likely to shop at these three stores. *Note:* While the software calculates this number to several places after the decimal point, you can round to the nearest whole person.

3. Close down the Facilities table.

4. Now in the Table Of Contents, right-click the Demand Points sublayer and choose Open Attribute Table. Take a screenshot for your instructor.

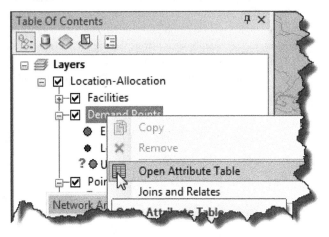

Notice in the Demand Points table that the Facility ID column is set to <Null> if the demand point is beyond the 5-minute cutoff point, but a numerical value is shown. It represents the ID of a facility to which the demand point was allocated.

The Weight column shows the population count that was loaded from the census tract feature class. The Allocated-Weight column illustrates the amount of demand allocated to the associated facility. The amount of weight allocated is based on the linear distance decay and the 5-minute cutoff parameters you set.

5. In the Table Of Contents, right-click the Lines sublayer and choose Open Attribute Table.

That table contains a single record for each demand point allocated to the facility and lists the shortest impedance between the two locations and the weight captured by the facility.

6. Quit ArcMap.

ADDITIONAL READING AND RESOURCES

Buster, J. Allison. *Designing Geodatabases for Transportation*. Redlands, CA: Esri Press, 2008.

Lang, Laura. *Transportation GIS*. Redlands, CA: Esri Press, 1999 (Multimedia CD including 12 case studies).

Scott, Allen J. "Location-Allocation Systems: A Review." *Geographical Analysis*, 1970.

KEY TERMS

allocation: In network analysis, the process of assigning edges and junctions to features until the feature's capacity or limit of impedance (e.g., number of students in a bus) is achieved.

connectivity: In network systems, the state of association between edges and junctions in a network used for modeling.

edge: In network systems, a special line feature through which traffic, substances, or other resources flow.

geometric network: A collection of interconnected edge and junction features that represent a linear network of nontransportation phenomena such as utilities or hydrological systems.

impedance: In network modeling, a number of conditions (such as speed limit or electrical resistance) that slow the movement along edges.

junction: In network modeling, the point where two or more edges meet.

location: In the context of location-allocation, modeling is the determination of the best place to put a facility depending on selected factors to be optimized, such as the shortest amount of travel time to the facility.

transportation network: Nonstandard terminology for a network dataset specifically designed for modeling the movement of traffic.

traversability: The degree to which a network can be used to accommodate movement from one place to another.

turn: In network analysis, a specialized feature class that defines turn movements in the ArcGIS software.

CHAPTER 8

Surface Analysis

LEARNING OBJECTIVES

Here is the content you will learn in this chapter:

1. Explain what a statistical surface is and what geostatistics is.

2. Explain the process of interpolation and describe the various methods of performing this process.

3. Explain the difference between the Geostatistical Analyst toolbox and the Geostatistical Wizard.

4. Define and explain what a semivariogram is and how it is used in nonlinear interpolation.

5. Describe what slope and aspect analysis is and how it might be applied.

6. Explain what spatial autocorrelation is and what it tells us about surfaces.

7. Explain what viewshed analysis is and why it is used.

8. Describe the idea of Cut-and-Fill and generally how it is done.

9. Use ArcGIS software to create a surface and perform interpolation.

10. Use ArcGIS to explore the output of spatial interpolation.

BEHAVIORAL INDICATORS

When you are finished with this chapter, you will be able to:

1. In your own words, define *statistical surface* and *geostatistics*.

2. In your own words, explain the process of interpolation and describe at least two different types.

3. Explain the difference between the Geostatistical Analyst toolbox and the Geostatistical Wizard.

4. In your own words, define *semivariogram* and explain how it is used in nonlinear interpolation.

5. In your own words, describe what slope and aspect analysis is and how it might be applied.

6. Explain what spatial autocorrelation is and what it tells us about surfaces.

7. Explain what viewshed analysis is and why it is used.

8. In your own words, describe the idea of Cut-and-Fill, its use, and generally how it is done.

9. Demonstrate, via artifacts, your use of the ArcGIS software to create a surface and perform interpolation.

10. Demonstrate, via artifacts, your use of the ArcGIS software to explore the output of spatial interpolation.

Chapter Overview

This chapter introduces you to a few of the more common tools associated with surface creation, characterization, and use. The first section discusses a set of techniques generally described as *geostatistics*. These are parts of a series of techniques associated with the interpolation and statistical characterization of the height values of statistical surfaces—those surfaces that are continuous and contain height values measured at the ordinal, interval, or ratio scales. You will get a chance to perform a Kriging interpolation based on default settings so you will become comfortable with the techniques. These techniques also include a few of the tools needed to statistically analyze the surface you are creating so it is as realistic a model as possible.

In addition to geostatistical analysis, you will learn a bit about the raster surface toolset and how it can be applied. In particular, you will learn a bit about slope, aspect, curvature, contours, contours with barriers, contour lists, hillshade, and Cut-and-Fill technique. Finally, you will get a brief introduction to visibility analysis and viewshed analysis, both based on the 3D Analyst extension. In all cases, you will get a chance to show your mastery of this material and your ability to find more information online and even to find meaningful online video tutorials to extend your knowledge.

Geostatistics (Interpolation)

Within the ArcGIS software, there is the Geostatistics extension. This term is often misunderstood to mean spatial statistics. While it does deal with statistics, **geostatistics** focuses on the analysis, characterization, and interpolation (prediction of points along a surface) of **statistical surfaces**. Statistical surfaces are any continuous surfaces composed of height (z), values measured at the ordinal, interval, or ratio scales. In short, statistical surfaces are continuous surfaces measured based on measurable height values. The values are often considered to represent topographic surfaces as measured by individual elevation heights, but there are many other statistical surface datasets that can be included. Examples of such datasets include temperature, rainfall, barometric pressure, and wind speeds.

Among the most important tasks of the geostatistics extension is to assist in the creation of these surfaces based on samples of z-values through one of a number of methods of interpolation. Most of you are familiar with linear interpolation where the assumption is that the surface changes in an arithmetic progression such as 10, 20, 30, 40, and so on. In such a system, prediction of missing values is simply a process of mathematically continuing the arithmetic progress. So, for example, a set of numbers like 20, 30, ??, 50, 60, and so on, where the ?? indicates the missing value, would simply involve mathematically determining the missing value by recognizing that the sequence progresses in units of 10. Any number halfway between 30 and 50 would be discovered by adding 30 to 50 and dividing by 2, resulting in a value of 40. Such progressions can be envisioned to represent vertical surface values rather than simply mathematical progressions. In the previous example, you were predicting that the missing value of 40 is an elevation value of 40. In this way, you can take a collection of sampled elevation values and predict a large collection of missing values and eventually construct lines of equal elevation known as *contour lines*.

This concept of interpolation, while elegant, is not very representative of how real statistical surfaces work. Think about how many hills or mountains you have seen that change in elevation in a perfectly linear fashion. This is quite rare and likely happens when they have been created by humans. Instead, topographic and other statistical surfaces will change in rather irregular progressions. Still, while irregular, there is one property each surface possesses that is universal: The closer any two values are to each other, the more likely they are to be very similar in z-value. Alternatively, the farther away any two points are from each other, the more likely they are to be unlike each other. This is a form of what statisticians call **spatial autocorrelation**. This process is used for a range of nonlinear interpolation methods that recognize the relationship between sample points and distance.

One fairly common approach to interpolation that takes advantage of these properties is called **inverse distance weighted** (IDW) interpolation. This method assumes that the farther away two sample points are, the less likely they are to be similar. Thus, when the interpolation takes place, when comparing z-values, the interpolation algorithm assigns less weight to sample points that are farther away from each other and assigns more weight to sample points that are closer to each other. This approach is considerably more accurate than the linear technique because it does not have to rely on the assumption of arithmetic progression, which you have seen is not very common.

Another nonlinear interpolation makes the same assumptions of nonlinearity of the surface and spatial autocorrelation but goes one step further. At some point, the degree of spatial autocorrelation diminishes to the point of having virtually no impact on the interpolation algorithm. This approach to interpolation is called **Kriging** and creates a model, called a **semivariogram**, that represents, based on a set of sample points, the relationship between the sample distance, called the *lag,* and the degree of similarity measured as a statistical semivariance (a measure of similarity). The semivariogram is represented as a graph with the lag (distance between samples) on the x-axis and the variance on the y-axis. As one moves along the x-axis from the origin, the variance increases rather quickly at first, slows, and eventually achieves a very shallow slope approximating the horizontal. At the point where this happens—that is, where the semivariance approaches a relatively unchanging state—is called the *sill*. The software uses this semivariance to determine the point at which it no longer takes sample distance into account when calculating interpolation.

ACTIVITY 8-1   INTERPOLATION

This activity will give you a chance to demonstrate your command of the material on interpolation and to learn still more through WebQuesting.

1. In your own words, define *interpolation*.

2. Do a search of the Internet to compile a list of at least five different types of interpolation. Next to each, provide a brief description of how it operates. *Note:* For extra credit, your instructor might suggest you also include when you might choose one method of interpolation over another.

3. Use the ArcGIS help menu and any other sources you choose to explain the difference between the Geostatistical Analyst toolbox and the Geostatistical Wizard.

4. Do a web search for the term *semivariogram*. Provide a quick sketch of such a diagram, label it, and explain how it is used in the process of Kriging.

5. Given that inverse distance weighting (IDW) and Kriging both rely on spatial autocorrelation, what is the difference between the two? Discuss briefly what Kriging adds to the process and how it might impact the results of the interpolation process.

ARCGIS LESSON 8-1 | CREATING A SURFACE USING DEFAULT PARAMETERS

This short lesson gives you the opportunity to generate a continuous surface based on the ArcGIS default parameters.

Part I: Activating the Geostatistical Analyst Extension and Add the Toolbar

Steps

1. Start ArcMap.
2. On the ArcMap Getting Started dialog box, click Cancel (assuming you have previously opened map data).
3. On the main menu, click Customize > Extensions.
4. Check the Geostatistical Analyst checkbox to activate it.
5. Click Close to shut down.
6. On the main menu, select Customize > Toolbars > Geostatistical Analyst. This adds the toolbar to your session and will remain the next time you open ArcMap.

Part II: Adding Data to Your Session and Saving the Results

Now you will add data to ArcMap and change its symbology to make it easier to understand.

Steps

1. On the Standard toolbar, click the Add Data button ✛.
2. Now navigate to the Geostatistical Analyst tutorial data (default location C:\ArcGIS\ArcTutor\Geostatistical Analyst).
3. Double-click the ca_ozone.gdb geodatabase to expose its contents.
4. Press the CTRL key and choose the 03_Sep06_3pm and ca_outline datasets.
5. Click Add.

6. Right-click the ca_outline layer legend (the box below the layer's name) in the Table Of Contents and click No Color (see figure).

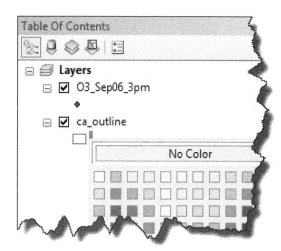

Only the California outline is displayed. This will allow you to see the layers as you create beneath them in this tutorial.

7. Double-click the 03_Sep06_3pm layer's name in the Table Of Contents.
8. In the Layer Properties dialog box, click the Symbology tab.
9. In the Show box, select Quantiles (http://tinyurl.com/q5gz35w) and click Graduated colors.
10. In the Fields box, set the Value to OZONE.
11. Select the White to Black color ramp from the pulldown so the point symbols will stand out against the color surfaces you are going to create in the tutorial.

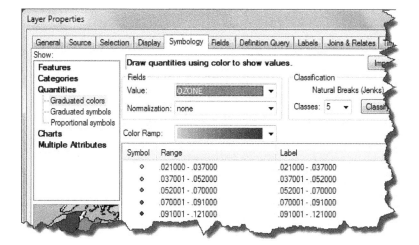

(continued)

ARCGIS LESSON 8-1 | (CONTINUED)

12. Click OK.

13. Look at the distribution and note that the highest ozone values occurred in the Central Valley and the lowest occurred along the coast. With the simple act of mapping the ozone, you are beginning the exploration of the phenomena. This will lead to greater understanding of the data you are going to model.

14. Now save your map by clicking File > Save on the main menu.

15. Browse to your working folder (the one you have created, e.g., C:\Geostatistical Analyst Tutorial). Make a note of where you created this because you will need it later.

16. In the File name text box, type "Ozone Prediction Map.mxd."

17. Click Save. Because this is the first time you save your map, you have to give it a name. Later on you'll only have to click Save.

Part III: Creating a Surface Using Defaults

Now you are going to use interpolation to create a surface from the sample ozone_concentration data using the default Geostatistical Analyst settings. You're going to use the (03_Sep06_3pm) point dataset for input and use ordinary Kriging to interpolate the ozone values at the locations where ozone values are not known. You are going to accept the defaults on many of the dialog boxes.

The primary purpose of this exercise is to get you used to the Geostatistical Wizard so you won't have to concentrate on the details. Still, it's a good idea to pause and look at each dialog box so you gain familiarity.

Steps

1. Click on the Geostatistical Analyst arrow on the Geostatistical Analyst toolbar and click Geostatistical Wizard to fire it up.

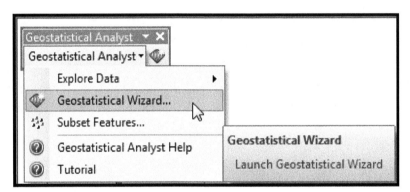

2. Click Kriging/CoKriging in the Methods list box.

3. Click the Source Dataset arrow and select 03_Sep06_3pm.

4. Click on the Data Field arrow and select the OZONE attribute.

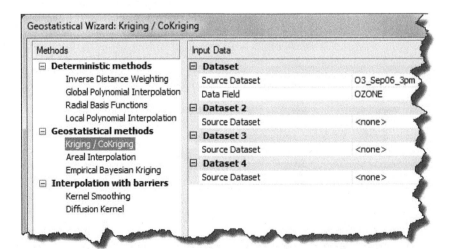

5. Click Next. The defaults Ordinary (Kriging Type) and Prediction (Output Surface Type) will be selected on the dialog box (if not, select them). Because the method to map the ozone surface has been selected, you can now click

Finish to create the default surface. The next five steps will show you additional dialog boxes. You can resize and drag the internal panels (windows) by moving the dividers separating them.

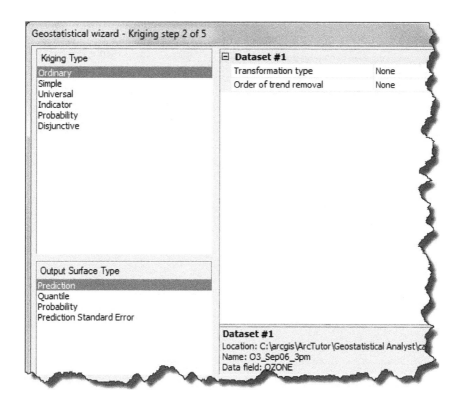

Note that the box on the bottom-right of the Geostatistical wizard in the preceding figure shows an abbreviated description of the highlighted method or parameter depending on

the context. Here it shows the dataset and the Data field you will use to create the surface.

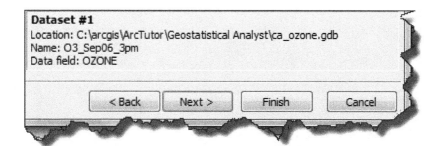

6. Click Next. The semivariogram/covariance model will be displayed, allowing you to see spatial relationships among measured points. As you saw earlier, the closer things are, the more alike they are. The semivariogram allows you to explore

that assumption. The process of fitting the semivariogram model to capture the spatial relationships you need for interpolation is called variography.

(continued)

ARCGIS LESSON 8-1 | (CONTINUED)

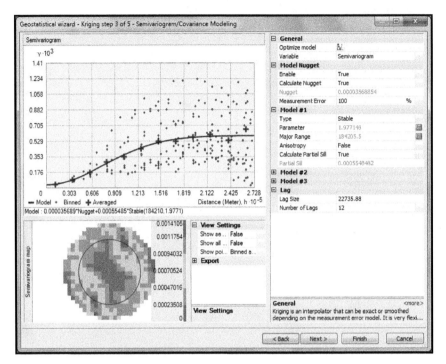

7. Click Next. The crosshairs show a location that has no measured value. To predict (interpolate) a value for that location, you can use the values at the measured locations for comparison. Recall that you know that the closer measured values are most likely to be like the unmeasured locations you are trying to predict. The red points in the preceding image are going to be weighted (influence the unknown values) more than the green points because they are close to the location you are predicting. Using the surrounding points and the semivariogram/covariance model, you can predict values for the unmeasured location.

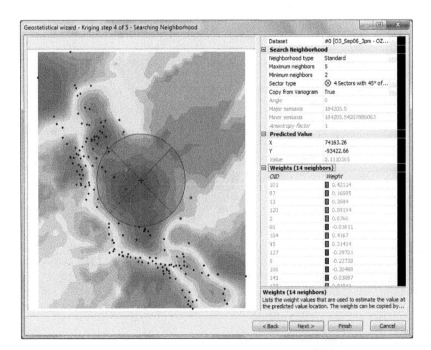

8. Click Next. The cross-validation diagram in the following
   image gives you an idea of how well the model predicts
   the unknown values.

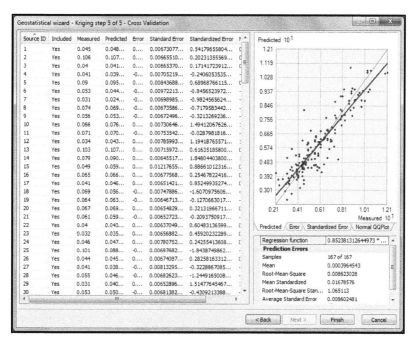

9. Click Finish. The Method Report dialog box summarizes the
   information on the method (and its associated parameters)
   that you will use to create the output surface.

(continued)

ARCGIS LESSON 8-1 | (CONTINUED)

10. Click OK. This adds the predicted ozone map to the top layer of the Table Of Contents.

11. Double-click the layer in the Table Of Contents to open the Layer Properties dialog box.

12. Click the General tab and change the name of the latter to "Default Kriging" to indicate you used default settings.

13. Click OK.

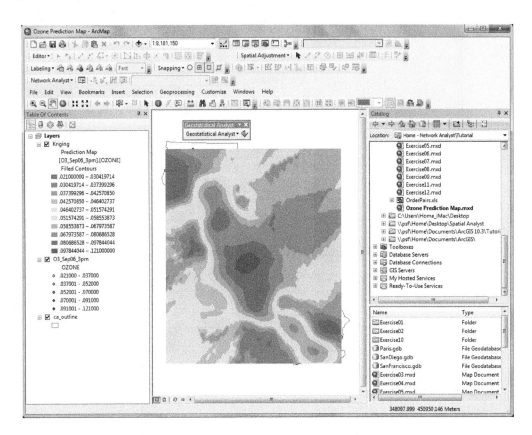

14. Click the Save button 💾 on the Standard toolbar to save your map.

    Note that the interpolations move onto the ocean because the extent of the layer is the same as the extent of the input data (03_Sep06_3pm).

15. To restrict the prediction surface to areas within the state of California, right-click the Default Kriging layer and click Properties.

16. Click the Extent tab.

17. Click the Set the extent to arrow, click the rectangular extent of the ca_outline, and then click OK. The interpolated area now extends to cover all of California.

18. Right-click the Layers data frame on the Table Of Contents, click Properties, and then click the Data Frame tab.

19. Click the Clip Options arrow, select Clip to shape, and then click the Specify Shape button.

20. On the Data Frame Clipping dialog box, click the Outline of Features button, and click the Layer arrow.

21. Now click ca_outline.

22. Click OK, and then click OK again. The predicted surface is now clipped so it only displays within the land areas of California. Take a screenshot of the following image for your instructor.

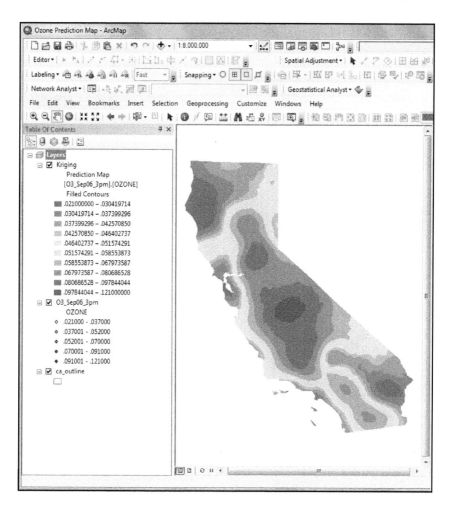

23. Drag the 03_Sep06 layer to the top of the Table Of Contents. Make a judgment about how well the Default Kriging layer represents the measured ozone values. Do high ozone predictions occur where the high ozone concentrations actually measured?

24. Right-click the Default Kriging layer in the Table Of Contents and click Validation/Prediction (see the following image). This will open the GA Layer To Points geoprocessing tool with the Default Kriging layer specified as the input geospatial layer.

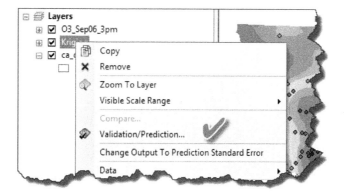

(continued)

ARCGIS LESSON 8-1 | (CONTINUED)

25. Check that Input geostatistical layer is set to default Kriging. For Point observations go to the geodatabase that contains your tutorial data and click the ca_cities dataset. In the following screen, leave the Field to validate on (optional) empty because you want to generate ozone predictions only for major cities. For Output statistics at point locations, navigate to the folder you created for the output and name the output "CA_cities_ozone.shp."

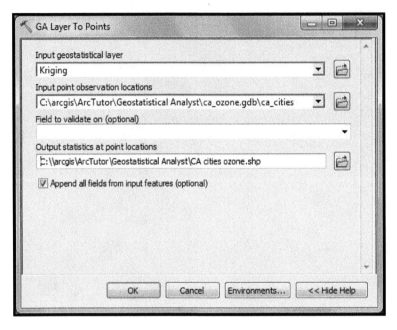

26. Click OK to run the tool.

27. Once the tool has run, click the Add Data button ✛ on the Standard toolbar.

28. Navigate to the data, click CA_ozone_cities_shape, and then click Add. The point observations layer is now added to your map.

29. Right-Click the CA_cities_ozone layer and click Open Attribute Table. Note that each city now has a predicted ozone value based on your interpolation, as well as a standard error value (the level of uncertainty associated with the ozone prediction model for each city).

30. Take a screenshot for your instructor.

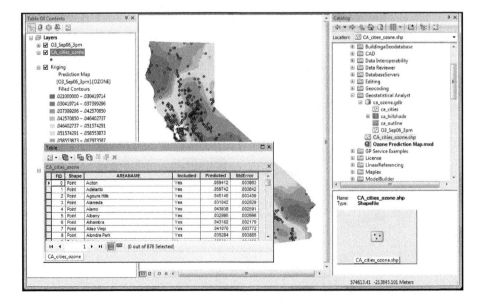

31. Close the Table window.

32. Right-click the CA_cities_ozone layer and click Remove to remove the layer from the project.

33. Save your work and log out.

Here is the process you just went through.

Represent Data → Explore Data → Fit Model → Perform Diagnostics
→ Compare Models

This is the standard approach to geostatistical analysis you will use whenever you perform this type of analysis. You might want to keep it in mind whenever you need to do an interpolation.

ARCGIS LESSON 8-2 | EXPLORING YOUR DATA

This exercise allows you to examine the surface you just created. In time, you will learn more about how to use the different approaches to surface analysis in practice, but for now it's important that you understand the nature of surfaces in general. You do this because to make better decisions when creating your surfaces, you should first explore your dataset to understand things like obvious error and how the data are distributed, to search for trends and directional bias, and so on. You are going to examine the overall distribution in the steps in the following two parts.

Part I: Examining the Distribution of Your Data Using the Histogram

While it might be nice to just go ahead and generate a surface, the best results for interpolation occur when the data are normally distributed (in the shape of a bell curve). If your data are skewed (lopsided), you might decide to transform the data to make them normal. To do this, you must understand the distribution of your data using frequency histograms. The histogram plots frequency for the attributes in the data, enabling you to examine the univariate (single variable) distribution for each attribute in each dataset.

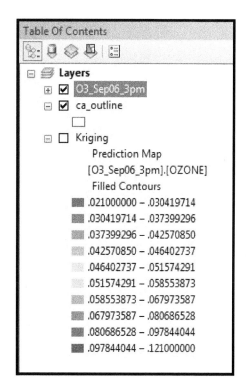

Steps

1. If you closed your previous exercise, begin again and open Ozone Prediction Map.mxd.

2. Click the ca_outline layer and drag it under the 03_Sep06_3pm layer in the Table Of Contents.

3. Click on 03_Sep06_3pm to select it.

4. On the Geospatial Analyst toolbar, click Geostatistical Analyst > Explore Data > Histogram.

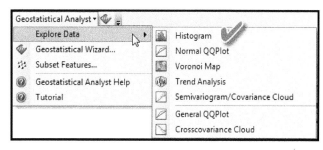

(*continued*)

ARCGIS LESSON 8-2 | (CONTINUED)

5. On the Histogram dialog box, click the Attribute arrow and select OZONE.

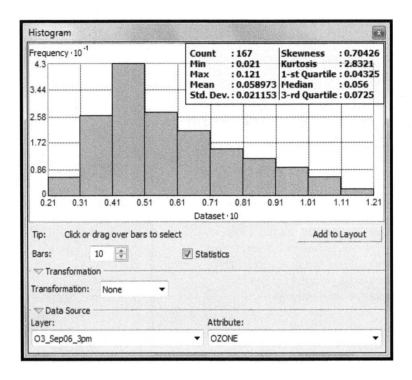

The x-axis numbers have been rescaled by a factor of 10 so it is easier to read. If needed, move the Histogram dialog box so you can see the map as well as the box.

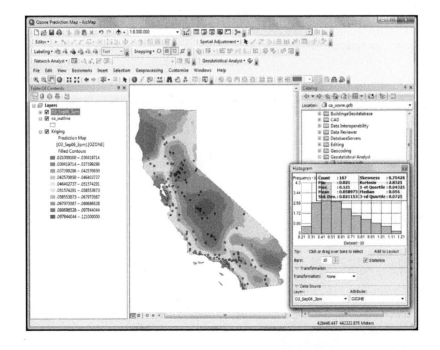

Take a look at the diagram you just created. Notice that ozone is shown in the histogram with the range of data grouped into 10 classes. The height of each bar represents the frequency (number of occurrences) for each of these 10 groups. The most common features occur in the middle and then spread out in the familiar bell shape. From basic statistics, you will see that the mean and median are roughly the same value, suggesting your data are normally distributed.

This distribution of ozone shows that the data are unimodal (a single lump) and slightly skewed (squeezed) to the right. The right tail shows the presence of a relatively low number of sample points with large concentration values, but overall, the distribution is still pretty close to normal.

6. Select the two histogram bars with ozone values over 0.10 parts per million (ppm). Keep in mind that the values have been rescaled by a factor of 10 for easy reading. Click and drag the pointer over these histogram bars. The sample points within the selected range will show on the map. Note that most of the sample points are in the Central Valley.

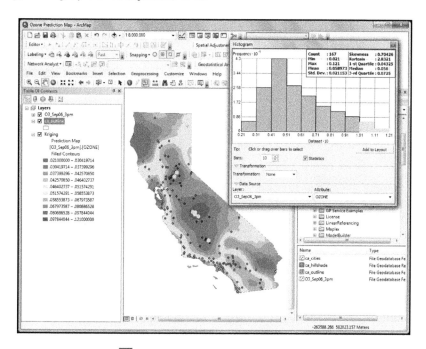

7. Now click the Clear Selected Features button on the Tools toolbar to clear the selected points on the map and the histogram.

8. Click the Close button located in the upper corner of the Histogram dialog box.

Part II: Creating a Normal QQPlot

A QQPlot (quantile-quantile) is used to compare the distribution of the data to what a standard normal distribution would look like.

This gives you another measure of distributional normality. To read this, realize that the closer the points are to the straight (45 degree) line in the graph, the closer the sample data are to a normal distribution.

Steps

1. Go to the Geostatistical Analyst toolbar, click Geostatistical Analyst > Explore Data > Normal QQPlot.

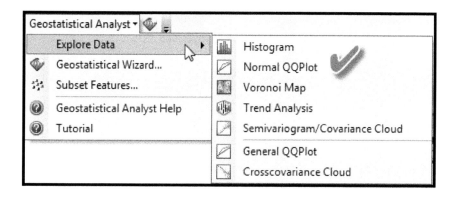

(continued)

ARCGIS LESSON 8-2 | (CONTINUED)

2. Click the Attribute arrow and select OZONE.

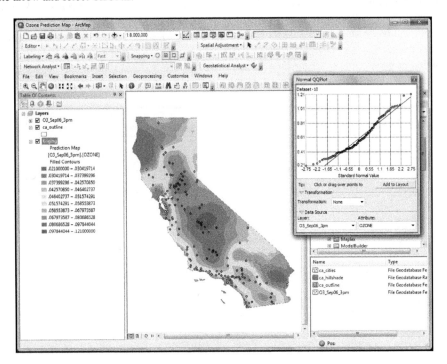

Notice that the quantiles from the two distributions are plotted against each other. If the two sets were identical, the distribution would be a straight line. In the distribution above, you notice that they are not a straight line. In fact, the primary difference occurs at low ozone concentration.

If the data do not exhibit a normal distribution in either the histogram or the QQPlot, you have a good case to perform some sort of data transformation to make it conform to the normal distribution before using selected Kriging interpolation techniques.

3. Click the Close button located in the upper corner of the Normal QQPlot dialog box.

Part III: Identifying Global Trends in Your Data

Another piece of useful information would be the general trend in the data. If the trend is a nonrandom (deterministic) part of your surface it can be represented by a mathematical formula. There are several landscape features such as glacial troughs (U-shaped valleys) or gently sloping hillsides that can be generally represented by such a formula, but even these shapes have portions that don't follow the general trend. The local variation can be added to a general surface model by modeling the trend first using one of these smoothing equations, removing it from the data, and continuing your analysis by modeling the differences (called *residuals*). In this approach, you are analyzing short-term (local) variation in the surface. Finally, the Trend Analysis tool allows you to identify both the presence and shape of the trend.

Steps

1. On the Geostatistical Analyst toolbar, click Geostatistical Analyst > Explore > Data > Trend Analysis.

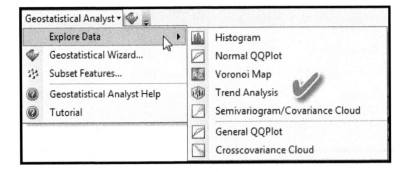

Now click the Attribute arrow and choose OZONE. Each of the vertical lines in the Trend Analysis plot represents the location and value (z-value) of each of the ozone measurements. The data points are projected onto the perpendicular planes as well as east–west and north–south planes. The software draws a line of best fit (a polynomial) through the projected points, showing trends in multiple specific directions. If the resulting line were flat, this would indicate that there is no trend. Note the green line in the following graphic. It starts out with low values, increases as it moves toward the center of the x-axis, and then decreases. Also note the blue line that increases as it moves north and decreases from the center of the state. What this shows is that the data seem to exhibit a strong trend from the center in all directions. Take a screen shot for your instructor.

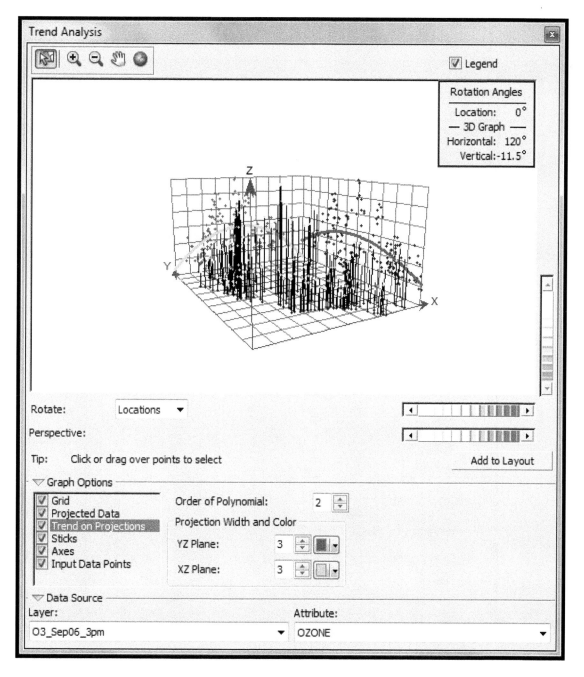

2. Now you get a chance to move the graphic around. Click the Rotate Locations scroll bar and scroll left until the rotation angle is 90°.

(*continued*)

ARCGIS LESSON 8-2 | (CONTINUED)

3. Notice that while you rotate the points, the trends always exhibit upside-down U shapes. Also, the trend doesn't seem to be stronger (meaning a more pronounced U shape) for any particular rotation angle, reaffirming the center of the data in all directions. Because the trend is U shaped, a second-order polynomial is a good choice to use as a global trend model.

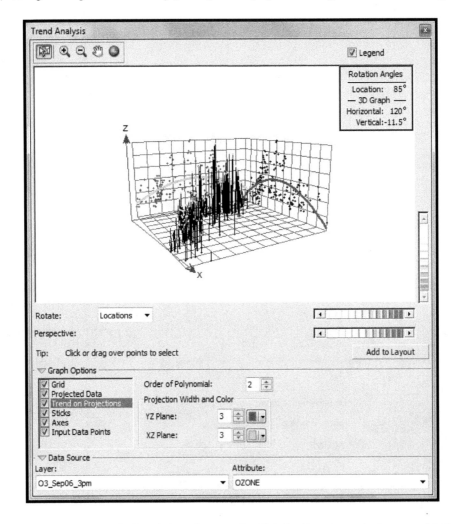

This gives you a nice overview of a few of the tools you can use to examine your surface data. Take a screenshot for your instructor. When you have finished, you might want to examine some of the other techniques. Before you quit, make sure to perform the next two steps.

4. Click the Close button in the upper right-hand corner of the Trend Analysis dialog box.

5. Shut down.

Raster Surface Toolset

Surface data are often represented as raster datasets as opposed to the **triangulated irregular network** vector model. This raster data model provides some very powerful tools for analysis quite different from the Geostatistical Analyst extension. The raster surface tools are available with the 3D Analyst toolset and begin with the raster surface dataset. Among surface properties and analyses for which you can use the tools include the following:

Aspect. Downslope direction of the steepest portion a surface is facing.

Contour. Creates a line feature class of contours from a raster surface.

Contour list. Creates a line feature class of selected contour values from a raster surface.

Contour with barriers. Creates contours from a raster surface with the inclusion of barrier features so you can create contours on either side of the barrier.

Curvature. A measure of the curvature of a surface including topographic profiles.

Cut Fill. A set of tools that calculates the difference in volume between two surfaces.

Hillshade. Creates a shaded relief visualization from a surface raster based on the source and angle of illumination (usually the sun).

Slope. A companion of aspect that finds the rate of maximum change in z-value from each cell within a raster surface. Recall that Aspect requires Slope before it can be calculated.

The following paragraphs will describe, in general, how each of these analyses works and how they might be used to create specific features from surfaces. Work with the ArcGIS 3D Analyst is normally an advanced technique, so you will learn these approaches primarily from a conceptual framework to inform you so that you will be prepared for any future 3-D work. This discussion proceeds alphabetically beginning with aspect.

Aspect

Aspect begins by calculating the maximum slope (covered subsequently) and measuring it clockwise in compass directions from 0° and 360° as direct north. As you can easily tell, east would be 90°, south 180°, and west 270°. Areas that are flat—by that I mean they have no slope—are assigned a value of −1 so the computer knows they have no measurable slope. Calculations are made by fitting a flat plane through a 3 × 3 set of z-value neighborhood. Some examples of the use of aspect include finding north-facing slopes for best ski areas, south-facing slopes for situating solar panels, or flat areas for easy agricultural use or for creating an airport landing strip.

Contour

Contour lines are linear features that identify lines of equal elevation. They are visually the same as an analog topographic map. In this context, the *term* contour is often used to represent other continuous surfaces such as temperature, barometric pressure, humidity, and the like. The general term for these lines is more accurately *isolines* rather than contours because contours refer specifically to lines of equal elevation. Other specific terms for their surface isolines include *isotherms* for temperature and *isobars* for pressure. The ArcGIS software, however, uses the general term *contours* for simplicity rather than for accuracy. As these lines become closer together, they graphically represent steep slopes while those lines that are farther apart indicate more gradual slopes. This differential spacing provides a nice visualization of the surface data for quick interpretation.

Contour List and Contour with Barriers

These two techniques are essentially part of the contour tools. In the case of Contour List, one can define specific contours from the contours generated through raster contour creation. One way of looking at this might be to consider selecting only those contours that represent different zones of vegetation along the side of a mountain. An additional useful modification of contouring is the ability to place a barrier in a surface, called Contour with Barriers. This allows one to generate contours, for example, on either side of a linear border independent of the overall surface dataset. The reason for using barriers is to generate more detailed surfaces around obstacles than can be generated with the general method of contouring.

Curvature

The curvature algorithms are measures of the shape of the surface both as a surface and as a cross-sectional profile. The approach fits a surface through the nine cells of a 3×3 matrix. If the result of that calculation is positive, it indicates that the surface is convex (bulging outward) while a negative curvature indicates that the surface is upwardly concave. A value of 0 means that the surface is flat. When calculating a cross-sectional profile, a negative value indicates the surface is upwardly convex, and a positive profile indicates the surface is upwardly concave. As with the surface, a value of 0 means the surface is flat.

Cut Fill

Cut Fill is performed based on a comparison of two coincident raster surfaces—a before surface and an after surface. The idea is to compare a surface displaying areas with volumes of surface material before and after removal (digging). The technique compares the two surfaces, summarizing the cell-by-cell comparison as three possible outcomes: those that have had material removed, those to which material was added, and those that had no change. This technique can be used to find regions of erosion and deposition in river valleys, to calculate areas where surface material is removed or areas to be filled for construction, or to analyze strip mining overburdening.

Hillshade

Hillshade is a visualization tool that creates a hypothetical illumination of a surface by determining a set of illumination values for each cell in a raster. To do this, the software sets a position for a hypothetical light source (normally the sun) and calculates the illumination values of each cell in relation to its neighboring cells. The azimuth of the sun is measured clockwise in degrees from 0 to 360. The default for this algorithm is 315°. Altitude is the slope angle of the illumination above the horizon in degrees from 0 at the horizon to 90 at zenith. The default is 45°. Hillshading enhances the visualization of topographic surfaces and is particularly useful to graphically compare a layer to the topography using the built-in transparency function of ArcGIS.

Slope

By definition, *slope* is the maximum change in z-value for each cell in a raster surface. Another common terminology for *slope* is *rise* (change in height) over reach (horizontal distance). The calculation ranges from 0° to 90° or 0° to infinity for slope measured as a percent rise. In the case of percent, the slope of a flat surface is assumed to be 0 percent slope while a 45° surface is 100 percent slope. Slope is an essential parameter as the aspect is based on the steepest slope. For this reason, the approach for calculating aspect is most often called *slope/aspect analysis*, although aspect is not required. Slope is very useful for determining road construction locations, places to build homes, and developing ski slopes.

ACTIVITY 8-2   RASTER SURFACE TOOLSET

This activity will give you a chance to demonstrate your command of the material on the raster surface toolset.

1. What is the purpose of hillshade?

2. Describe, in your own words, the terms *contours, contour list,* and *contour with barriers.* Do a quick web search and find examples of how contour list and contour with barriers might be used. Include them here.

3. In your own words, describe what Cut-and-Fill is and how it can be used.

4. Why is slope/aspect used as a term rather than just slope and then aspect?

5. What is curvature? Describe what it is used for in your own words.

6. Do a web search and list at least five particular uses of Cut Fill.

7. Because you will not actually use ArcGIS for the raster surface toolset, perform a Google search limited to videos. View a number of examples on videos of the use of the raster surface toolset. Copy the URL of the videos you find and share them with your instructor.

Viewshed Analysis

A final tool that is based on surfaces is called **viewshed analysis** or sometimes visibility analysis. As you are aware, if you've gone outside, the terrain, combined with buildings, vegetation, and other obstructions, often determines what you can and cannot see from any particular vantage point. Such visibility is vital in combat situations where combatants either want to hide from their enemy or be able to view the activities of their enemy. Closer to home, the value of real estate may partially depend on the ability to see parks, ocean front, or mountain scenery. Riverboat tours depend on the ability of customers to see the sights along the river. All of these depend on a technique called *line of sight*.

Line of sight is a line between any two points that indicates whether or not they are visible or hidden from an observer. In practice, a line of sight extends over portions of terrain that vary from visible to obstructed. Viewshed analysis extends this line-of-sight approach by identifying the cells in an input raster that can be seen from one or more locations. Rather than a line, this technique creates for each cell in a raster surface a set of values indicating how many observer points can be seen from each location. If you have only a single observer point, each cell that the observer can see will receive a value of 1 and those that the observer can't see will be assigned a value of 0. Observer points feature class used can contain points (for point observation locations) or lines (for linear observation places like roads). The nodes and vertices of lines are used as observation points rather than the entire lines.

Without getting into too much detail, the viewshed tool allows you to modify your analysis by adding elevated observation points (e.g., the top of a building) and obstructions to the analysis using the Offset portion of the Observer Points tool. You can even evaluate the quality of what is observed by assigning it a quality and weight. There is, after all, quite a difference between the visibility of a scenic vista versus a city dump or a local sewage lagoon. Even more detail can be added to account for curvature of the earth and refraction corrections. A prerequisite of visibility analysis and viewshed analysis is the 3D Analyst package.

ACTIVITY 8-3   VIEWSHED ANALYSIS

This exercise will allow you to demonstrate your mastery of the basic information about visibility analysis and viewshed analysis.

1. In your own words, explain the difference between visibility analysis and viewshed analysis.

2. Do a web search of viewshed analysis and compile a list of five unique uses of the technique. Copy these URLs to turn in to your instructor.

3. In your own words, describe the purposes of the Observer Points tool inside viewshed analysis.

4. Search the web to identify three viewshed analysis tutorials. Rank each of these in terms of their quality for explaining the technique. Copy the URL for each and turn them in to your instructor.

ADDITIONAL READING AND RESOURCES

Childs, Colin. *Interpolating Surfaces in ArcGIS Spatial Analyst*, ArcUser. Redlands, CA: Esri, 2004. http://www.esri.com/news/arcuser/0704/files/interpolating.pdf [last visited October 15, 2015].

Leusen, M. van. "Viewshed and Cost Surface Analysis Using GIS (Cartographic Modeling in a Cell-Based GIS II)," in *New Techniques for Old Times. CAA98 Computer Applications and Quantitative Methods in Archaeology. Proceedings of the 26th Conference, Barcelona, March 1998 (BAR International Series 757)*, ed. J.A. Barcelo, I. Briz, and A. Vila. Oxford: Archaeopress, 1999, 215–24. http://proceedings.caaconference.org/files/1998/34_Leusen_CAA_1998.pdf [last visited October 15, 2015].

Moore, I.D., R.B. Grayson, and A.R. Landson. "Digital Terrain Modelling: A Review of Hydrological, Geomorphological, and Biological Applications." *Hydrological Processes* 5 (1991): 3–30.

Zeverbergen, L.W., and C.R. Thorne. "Quantitative Analysis of Land Surface Topography." *Earth Surface Processes and Landforms* 12 (1987): 47–56.

KEY TERMS

geostatistics: Statistical techniques applied to continuous surfaces to explore and characterize them or to allow for predicting missing values from a sample of points (interpolation).

inverse distance weighted: A method of interpolation in which the values used for interpolation are weighted in decreasing amounts based on the increasing distance from the immediate sample point.

Kriging: A method of interpolation in which the surrounding values are weighted based on the distance between the measured points, the predicted locations, and the overall spatial arrangement of measured points. It employs a semivariogram that describes the shape of the curve of that spatial arrangement.

semivariogram: A graph generated during the Kriging interpolation process that represents the variance in measure of the distance between all pairs of sample interpolated positions.

spatial autocorrelation: The degree to which a set of spatial features (e.g., sample locations) and their associated values (e.g., elevation values) tend to be clustered together in space (positively autocorrelated) or dispersed (negatively autocorrelated).

statistical surfaces: Any set of nominal, ordinal, or interval data that varies continuously over geographic space. Examples include temperature, barometric pressure, and elevation values.

triangulated irregular network: A vector representation of statistical surfaces by dividing geographic space into continuous nonoverlapping triangles. The vertices of each triangle are sample points representing x and y locations and z (elevation) values.

viewshed analysis: An analytical technique performed on elevational surfaces that uses multiple line-of-sight calculations to identify the locations visible from one or more viewing locations.

CHAPTER 9

Constructing Models

LEARNING OBJECTIVES

Here is the content you will learn in this chapter:

1. What a GIS model is.
2. The ModelBuilder user interface.
3. What iterators are and how they work.
4. What Model Only Tools are available and how they work.
5. How to execute the tools of ModelBuilder.
6. How to create tools with ModelBuilder.
7. How to create a model in ModelBuilder.
8. How to add tools and data to a model.
9. How to run a model.
10. How to save a model.

BEHAVIORAL INDICATORS

When you are finished with this chapter, you will be able to:

1. Define *GIS model.*
2. Describe where to find ModelBuilder and demonstrate how to use it.
3. Briefly describe the process of ModelBuilder iterators and what they are used for.
4. List and briefly describe the Model Only tools available for modeling.
5. Demonstrate, using ModelBuilder, the execution of at least one of the ModelBuilder tools.
6. Explain how ModelBuilder can create tools.
7. Demonstrate, using ModelBuilder, the creation of a model.
8. Demonstrate, using ModelBuilder, the addition of tools and data to a model.
9. Demonstrate, using ModelBuilder, how to run a GIS model.
10. Demonstrate how to save a model and explain why this is useful.

Chapter Overview

This chapter introduces you to the idea of complex GIS models. GIS models are created by small analytical steps, each combined in a systematic way to either describe in detail how various spatial components combine to create a given spatial situation or to simulate a process taking place on a portion of Earth. Models can be created using any combination of the many tools available in GIS, and these all rely on a thorough understanding of the underlying geography. Given their often complex nature, it is useful to be able to document how a model was constructed so that it can be rerun if changes need to be made or if the model is unsuitable. In the past, flowcharts were used to link the data layers through processes to achieve a result. Today, the ArcGIS toolkit contains a set of tools collectively called **ModelBuilder** whose purpose is to design visual models of geographic processes and then incorporate real data and analytical methods within the graphic flowchart itself. This allows the user to save the model, modify it or change related data, and rerun the model at will. This chapter provides an overview of how ModelBuilder works.

GIS Models and ModelBuilder

In very simple terms, ModelBuilder is an ArcGIS application or visual language that allows for the easy creation of GIS models by linking appropriately sequenced geoprocessing steps. The models that this application is used for can be very simple, consisting of a combination of only a couple of maps or a map and a variable. They are generally designed to answer a geographic question such as those proposed in Chapter 1, Section 2 (What GIS Does).

Unlike Network Analyst, Spatial Analyst, and similar toolboxes, ModelBuilder does not have separate toolboxes residing on a disk and available in ArcCatalog, ArcToolbox, or the Search window. Instead, it is integral to the ArcGIS package itself and is available by clicking the ModelBuilder icon on the standard ArcGIS menu (Figure 9-1). When you do that, a new menu pops up with all the tools you need (Figure 9-2).

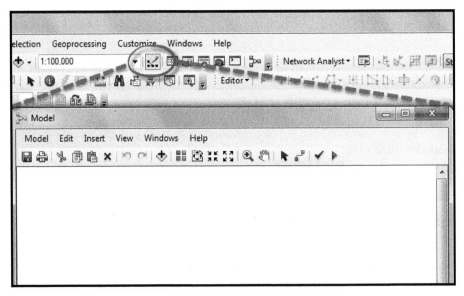

FIGURE 9-1 ModelBuilder icon location.

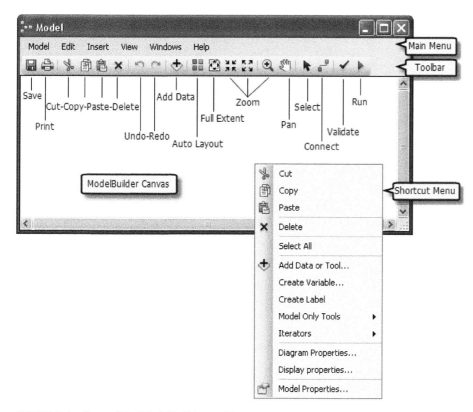

FIGURE 9-2 Parts of the ModelBuilder toolkit.

As you can see, there are quite a few options for ModelBuilder, giving it amazing power to use the available tools in ArcGIS to build quite sophisticated models. In fact, ModelBuilder can be used to create new tools to be applied to other models. It also has the power to work with the Python language interface in ArcGIS to make still more tools.

Beyond implementing once-only operations, ModelBuilder possesses tools to allow for very sophisticated model logic akin to traditional programming languages. Advanced Model-Builder models allow for the substitution of variables, use of list variables, iteration for repeating steps, feedback conditions that determine how a model process proceeds, preconditions determining tool execution, if-then-else logic, the ability to interact with the model as it is running, and the power to integrate other models, scripts, and external programs.

Iterators are just one example of the powerful functions of ModelBuilder as they allow a process to take place many times without the user having to stop and start the model each time. A good example of this is a model showing how the process of a forest fire might proceed over time. In a grid cell environment, each grid cell becomes increasingly susceptible to fire as it dries out from the heat of a fire in a cell next to it. Each time sequence adds to the likelihood that the target grid cell will catch fire. At a particular step in the process, it will then catch fire, and its neighbor grid cell must then be susceptible to the danger of fire. In the past, each step of the model had to be run independently, making the process tedious and time consuming. The other advanced tools you've seen described for ModelBuilder provide a simpler approach to processing tedious tasks and other complex modeling tasks to proceed independently of the user.

While ModelBuilder tools are beyond what can be covered in an introductory class, you should be aware of their availability and consider taking additional coursework or online tutorials to learn more. Meanwhile, review what you've learned and then proceed to the ArcGIS tutorial.

ACTIVITY 9-1   **GIS MODELS AND MODELBUILDER**

In this activity, you will get a chance to demonstrate your command of GIS models and the GIS ModelBuilder toolkit.

1. In your own words, explain what a GIS model is. Go on-line and provide at least five links to GIS models that use the ModelBuilder tool as their interface.

2. If ModelBuilder is not in the toolbox, isn't its own toolbar, and doesn't show up in the list of tools, explain why.

3. Label the following diagram and then prepare a list of what each icon does.

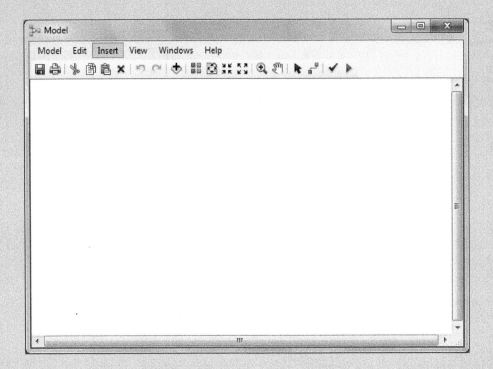

4. What are ModelBuilder iterators, and what are they used for? Search the Internet and find one example other than that in your textbook where iterators were used.

5. There are seven geoprocessing tools that support advanced ModelBuilder behavior and are not found in the tool dialogue box or in scripting. Go into your Help menu, find and list these seven tools, and describe each in your own words.

6. Your Help menu has a nice section called A quick tour of creating tools with ModelBuilder. Read this section and, in your own words, describe what it means to create tools with ModelBuilder. How does it differ from using tools?

Executing Tools in ModelBuilder

As you have learned, ModelBuilder allows you to both execute existing tools and create your own tools for later use. This section focuses just on the execution of existing tools using the ModelBuilder interface. GIS modeling can get quite complex, but you should be aware that most models in use today are simple, straightforward, and easy to understand. One of the advantages of the ModelBuilder interface is that the graphic model created by ModelBuilder and laid out on the model canvas provides a very nice way not only to conceptualize the model for yourself (Figure 9-3), but also to explain to clients how the model was built. Among its more powerful characteristics is also its ability to be easily modified and run without each layer's data being reloaded and each operation being requested each time. Instead, the interface allows the data to be included and the operations to be set so that they remain in place until modified.

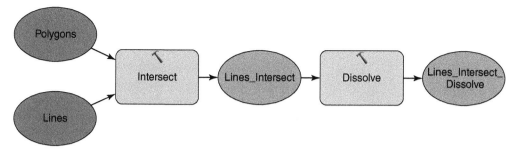

FIGURE 9-3   ModelBuilder flowchart example.

The next section provides you with an opportunity to create such a model, run it, and save it. The data for the model are included in your ArcGIS disk (or download), but to make sure you don't mess it up, it's a good idea to copy the data for use in the exercise rather than using the original data.

ARCGIS LESSON 9-1 | EXECUTING COMMANDS IN MODELBUILDER

This is a quick exercise of how to create, execute, and run a simple ModelBuilder model. You will gain experience using the ModelBuilder user interface, adding tools and data, and filling in necessary parameters. While you go through this process, pay particular attention to the user interface. The quicker you learn the interface, the more proficient you will become at the ModelBuilder process.

1. Start ArcMap.

2. Click the Open File (or Ctrl-O).

3. From within the window that opens, migrate to and select the ModelBuilder folder.

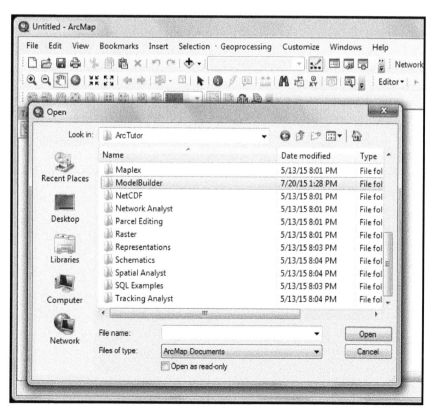

(continued)

ARCGIS LESSON 9-1 | (CONTINUED)

4. Select Extract Vegetation.mxd and click Open.

5. This opens the Extra Vegetation.mxd.

6. Click the ModelBuilder button on the ArcMap Standard tool-bar. This opens the ModelBuilder window.

7. You are now going to create a new "blank" model on the **ModelBuilder canvas**.

8. To build a model, you begin by adding tools and data. Start by clicking Geoprocessing > Search For Tools. This opens a search window that you can dock anywhere in ArcMap that you like. *Note:* When you click on the Geoprocessing menu, the buffer tool is displayed, but this cannot be placed on the canvas, so ignore it.

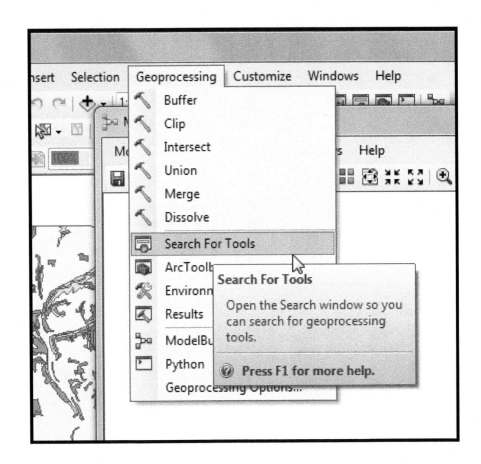

9. In the Search for Tools, type in the word "buffer." This opens a list of tools (with little hammers), among them the Buffer (Analysis) (Tool). Drag the tool onto the canvas.

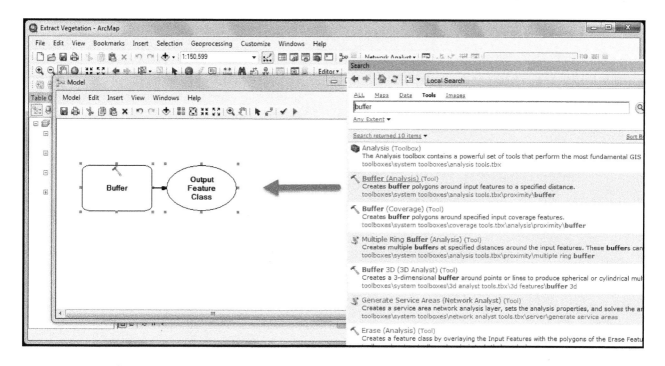

10. This adds an output variable (required) as well as the Buffer
     tool, both linked by a connector. Notice that they have no
     color because they don't have any parameters yet.

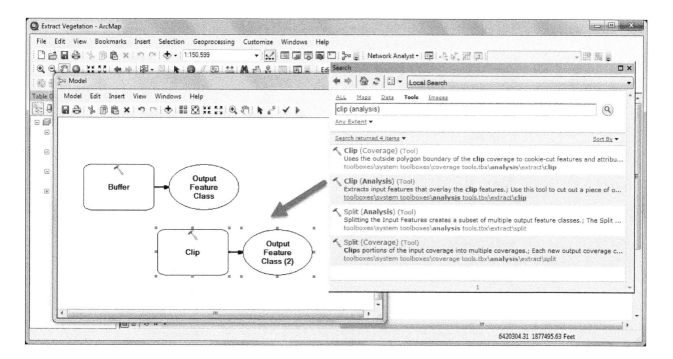

(*continued*)

ARCGIS LESSON 9-1 | (CONTINUED)

11. Click the Auto Layout button. This arranges the tools.

12. Now you need to fill in the parameters. Begin by double-clicking the Buffer tool on the canvas to open the dialog box.

13. For the Input Features parameter, click the Browse button and navigate to the input geodatabase (C:\ModelBuilder\ToolData\Input.gdb). From there choose the PlanA_Roads feature class and click Add. The Output Feature Class parameter is generated automatically.

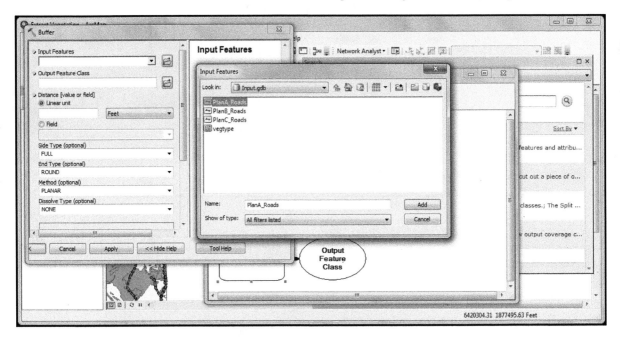

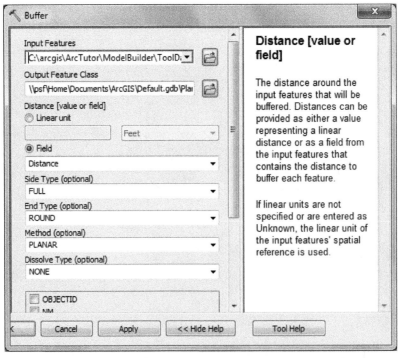

Replace this autogenerated name by clicking the Browse button for the Output Feature Class parameter. Navigate to the output geodatabase in the ModelBuilder folder (C:\ModelBuilder\Scratch\Output.gdb), type BufferedFC for the output name, and then click Save.

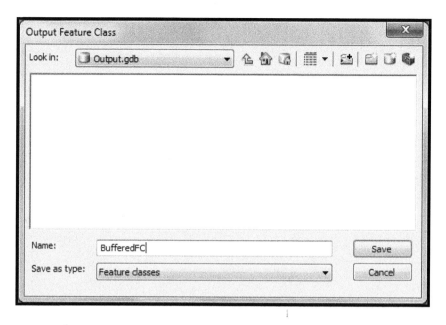

14. For the Distance parameter, choose the Field option and choose the Distance field from the drop-down list.

15. Click OK. (You don't need to fill in any other parameters.) Note how the modules now appear in color, indicating they have been populated.

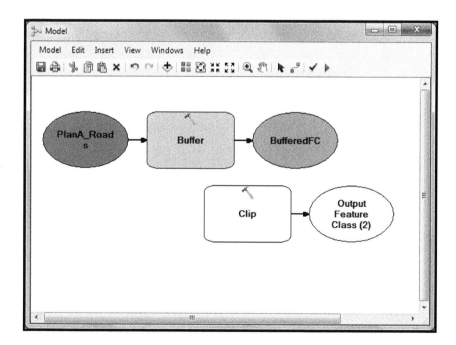

(*continued*)

ARCGIS LESSON 9-1 | (CONTINUED)

16. You also note that the Clip tool and Output Feature Class (2) do not appear in color, indicating they are not populated. To change that, double-click the Clip object on the canvas to open the tool's dialog box.

17. For the Input Features parameter, click the Browse button and navigate to the input database (C:\ModelBuilder\ToolData \Input.gdb).

18. Choose the "vegtype" feature class and click Add.

19. For the Clip Feature parameter, click the arrow and choose BufferedFC (recall you created this) from the drop-down list. The blue recycle symbol means that BufferedFC is a model variable. (See, you created a variable when you added the Buffer tool.)

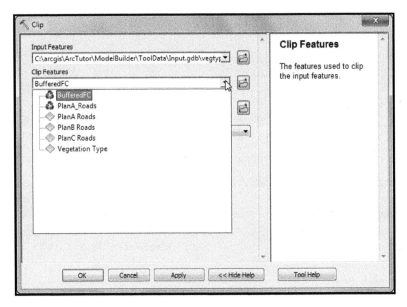

20. For the Output Feature Class, migrate to the Output.gdb again and in the Name location, type in "ClippedFC."

21. Notice that when you do that, the Clip objects populate *and* that the BufferedFC variable that you referenced is now linked to the Clip function to complete the model.

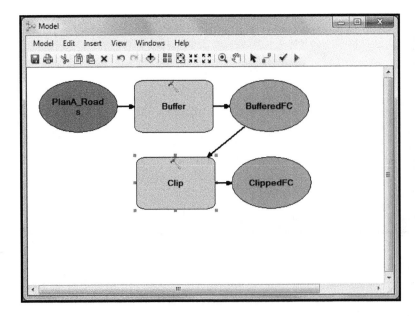

22. Ready to run the model? Hold on. Before you do that, right-click on the ClippedFC variable and check the Add to Display box. This means that the output will be added to the display in ArcMap.

23. From the Model menu, click Model > Run Entire Model.

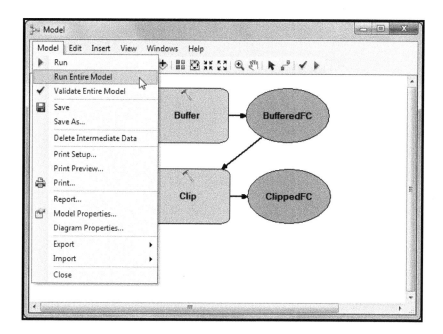

24. You will know that the model has run because as in the preceding screen, the tools (yellow rectangles) and the output variables (green ovals) have a gray drop shadow around them. When complete you will get a map showing the buffered and clipped roads as shown below.

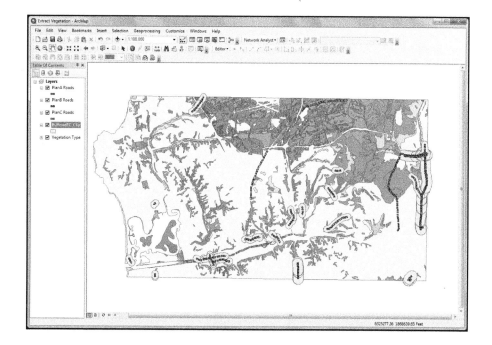

(*continued*)

ARCGIS LESSON 9-1 | (CONTINUED)

25. To save the model, click the Save tool on the ModelBuilder toolbar and navigate to C:\ ArcTutor\ModelBuilder.

26. A model can be saved only in a toolbox. Click the New Toolbox icon on the Navigation window. A toolbox with a default name is added in the workspace.

27. Select My Tools.tbx and click Save with the default name (Model).

28. If you want to try to run the model using a different set of roads (e.g., PlanB or PlanC feature classes), you can do one of the following:

   a. Double-click Buffer and navigate to another feature class, or

   b. Drag the data from the Catalog window onto the Model-Builder canvas to create a new data variable, and then connect this variable to Buffer.

29. Try one of these techniques out and return the results to your instructor. Provide a description of what the model did and why you buffered and clipped.

ARCGIS LESSON 9-2 | GETTING SUMMARY STATISTICS

This exercise will give you a chance to add the Summary Statistics tool to get a summary table of the affected area by vegetation type within the buffer polygons around the proposed roads.

1. As with ArcGIS Lesson 9-1, load the Extract Vegetation.mxd database.

2. Do a search for the Summary Statistics tool in the Search Window.

3. Drag the tool onto the ModelBuilder canvas.

4. Double-click the Summary Statistics element to open its dialog box.

5. From the Input Table parameter, click the input folder and choose ClippedFC with the blue recycle icon next to it from the list. This is your model variable.

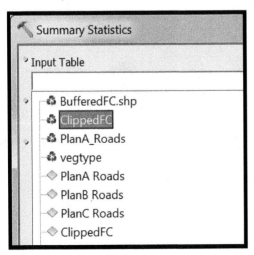

6. For the Output Table parameter, browse to the output geodatabase as before (C:\ModelBuilder\Scratch\Output.gdb), and give it the name AffectedVegetation and click Save.

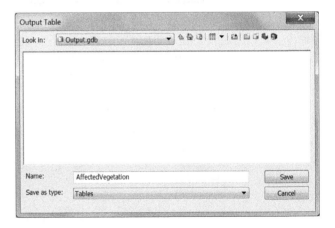

7. For the Statistic Field(s) parameter, select Shape_Area from the list.

8. Click on the cell to the right of Shape_Area under Statistic Type and choose SUM from the list.

9. Select VEG_TYPE for the Case field parameter and click OK. The completed dialog box follows.

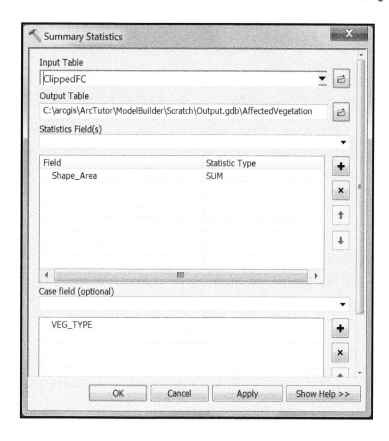

You should also notice in the following image that the model parameters are all filled in now.

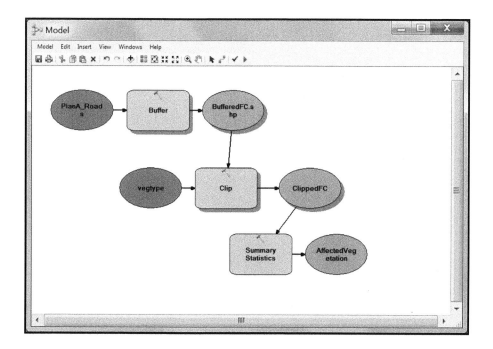

(*continued*)

ARCGIS LESSON 9-2 | (CONTINUED)

10. Be sure the Add to Display box is checked for the ClippedFC object by right-clicking it. If not, add it to the display now.

11. Run the model.

12. The output ClippedFC and the AffectedVegetation table will be added to the display in the ArcMap Table Of Contents.

13. Open the table by right-clicking and clicking Open. The table now shows a summary of area by vegetation type within the buffer polygons that will be affected by the propose roads for plan A.

14. Provide the final output to your instructor with a brief description of what the model was demonstrating in the real world. *Note:* The model is completely fictitious, but the idea is correct.

ADDITIONAL READING AND RESOURCES

Allen, D.W. *Getting to Know ArcGIS ModelBuilder.* Redlands, CA: Esri Press, 2011. http://desktop.arcgis.com/en/desktop/latest/analyze/modelbuilder/what-is-modelbuilder.htm

Mitchell, A. *The Esri Guide to GIS Analysis, Volume 3: Modeling Suitability, Movement, and Interaction*, 2012.

KEY TERMS

iterator: A ModelBuilder function that allows repetitive processes to take place without user intervention.
ModelBuilder: An ArcGIS application for all users to create and manage models by the use of an active visual flowcharting language that implements steps using real data.

ModelBuilder canvas: White space on which models are created.

CHAPTER 10

Reporting Results

LEARNING OBJECTIVES

Here is the content you will learn in this chapter:

1. What nonmap output is readily available in ArcGIS.

2. How nonmap output can be combined with maps to add information.

3. What route logs are and why they are useful.

4. How charts can be created.

5. How nonchart diagrams can be created.

6. How tables can be embedded into a map to add information.

7. How web maps can be created.

8. What story maps are, and why they are useful.

9. Some examples of different types of story maps.

10. How you use advanced story maps and features services.

BEHAVIORAL INDICATORS

When you are finished with this chapter, you will be able to:

1. List nonmap output from ArcGIS.

2. Use the software to embed nonmap output in a map layout to add to the information conveyed by the map.

3. Define what a road log is, and then, using a road network, perform a shortest path analysis and produce a road log as part of the output.

4. Create charts from existing cartographic tabular data.

5. Create nonchart diagrams from existing cartographic tabular data.

6. Demonstrate embedding a chart and other diagrams in an existing map layout.

7. Use ArcGIS online to create a simple web map.

8. Define what a story map is and why it is useful.

9. List several types of story maps and demonstrate how to create at least two different types using ArcGIS online.

10. Define *feature service*, explain why you might want one, and demonstrate how to create a simple one.

Chapter Overview

In the past, the vast majority of output from GIS analysis was composed of maps and maplike documents. Traditionally, the topic of creating maps came under the purview of cartographers and was thus taught either in a cartography class or a map use and/or reading class. Today many maps are still produced from GIS, but the software continues to offer an increasing array of noncartographic products. Some of these products are graphic devices, such as **histograms**, **bar graphs**, **line graphs**, and **pie charts**, with which you are already familiar. Others that you also have experienced but don't often associate with GIS are **route logs** that you see when you use such rudimentary GIS software as Yahoo Maps and Google Maps. There are even more exotic capabilities available such as animations and fly-throughs that give the observer much better ability to observe the results than with a static map.

All of these graphic devices can be created independently of each other although they typically use the same attribute tables. What is really powerful is the ability to embed tables, charts, diagrams, road logs, and other information on the layouts of any map document. You can even embed one map in the layout of another map to get a different view of a portion of the study area—for example, a blowup of a small section of a map to provide detail while the background map maintains the context and area. In this chapter, you will get a chance to work with some of these output types. Take some time to experiment with some of the others as well.

Charts and Diagrams

While maps are a very effective means of communicating information, many users are not as familiar with the subtleties of map construction and map interpretation as they are with things like the tables, charts, and diagrams they experience using office products like Excel. Fortunately, the same capabilities exist within ArcGIS so that it is able to produce bar charts, histograms, bar minimum and maximum charts, line and area charts, scatter plots, bubble plots, polar plots, pie charts, and scatterplot matrix diagrams. These diagrams can be produced using data from feature attributes, raster integer values, and nonspatial data from dBASE (.dbf) files and Excel tables. To make tables, you need numerical values. Text fields are available for labeling but not for creating the actual charts themselves.

ACTIVITY 10-1 **NONCARTOGRAPHIC GIS OUTPUT**

To make things simple, I'm going to categorize two different types of noncartographic output: visual (graphic) and nonvisual (statistical). As you will see, the line between purely visual and purely statistical is not well defined. Some graphics are designed to display the statistical data located in attribute tables. Other graphics are very advanced such as those designed to display, for example, statistical surfaces like topography either as single-frame observations or "fly-throughs." These techniques are useful but a bit beyond an entry level of GIS and of use in ArcGIS add-ons such as 3D Analyst. Other techniques involve the display of time-related spatial data, such as the animation of landuse change through time or the display of crime incidents through time. Such animation techniques, as with the fly-through approaches, provide us with a lifelike view of changing features through time. These animations are much like viewing television or movies in that they tend to be much more engaging than a traditional single map. As your GIS skills increase, you will learn to create such animations as you need them. You will get an opportunity to find examples of these techniques to whet your appetite. Meanwhile, I will introduce you to some of the simpler and typically more commonplace approaches to noncartographic output.

1. Go online and do a search for the term *GIS fly-through*. Try to find some that are primarily related to topography and a few others that are related to buildings/urban areas.

   a. List at least three URLs related to topographic fly-throughs.

   b. List at least three URLs related to nontopographic environments (e.g., urban areas).

   c. Describe your experience viewing these fly-throughs. For example, what did you see that you might not have been able to see with just a flat map? Did you find the fly-through engaging or distracting, clunky or elegant, useful or useless?

ARCGIS LESSON 10-1 | CREATING GRAPHS IN MAP LAYOUT

This is an example of how to create a graph and embed it within a map layout.

1. Open ArcMAP.
2. Navigate to the Editing folder from within the ArcTutor folders.
3. Select the Zion geodatabase.
4. From within that database, select the Streams and Boundaries layers and any other layer you might wish to experiment with.
5. Highlight the Streams layer, go to the selection pull-down menu on top, and choose Select By Attributes.
6. When the new menu appears, search in the streams layer for Type = 'perennial'. Notice that only the perennial streams appear highlighted. Now you want to see how the length of these perennial streams compare by length in kilometers.

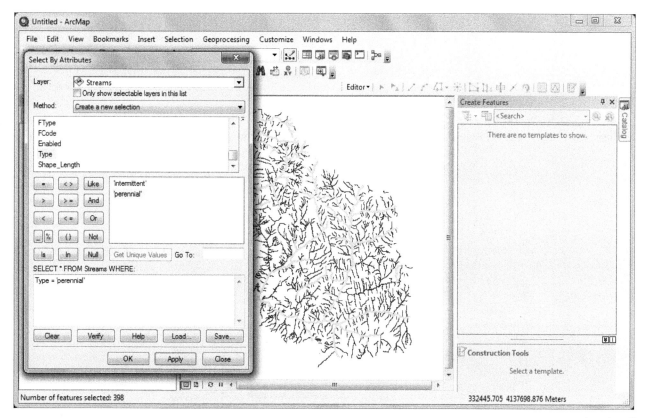

(continued)

ARCGIS LESSON 10-1 | (CONTINUED)

7. Highlight the streams layer again and go to the View pull-down menu on the top of the screen and select Graphs and then Create Graph.

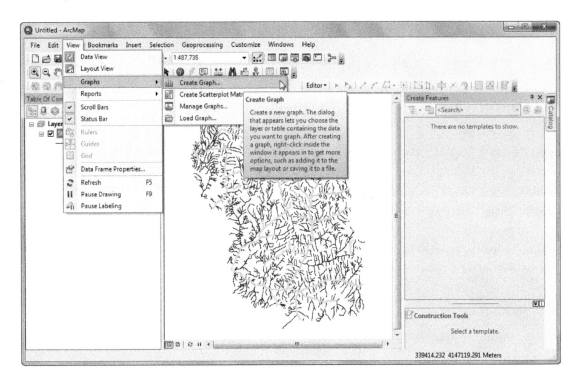

8. The Create Graph Wizard appears. Select Vertical Bar as the graph type, LengthKM as the Value Field you are searching, and GNIS_Name for the X field (optional) and for the X label field.

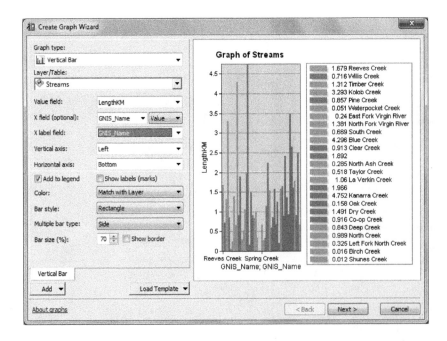

9. Notice that the legend shows both those streams you selected "perennial" and those you did not select "intermittent." To change this, click the Next button, and select the radio dial on the upper left that says "show only selected features / records on the graph."

10. Click Finish.

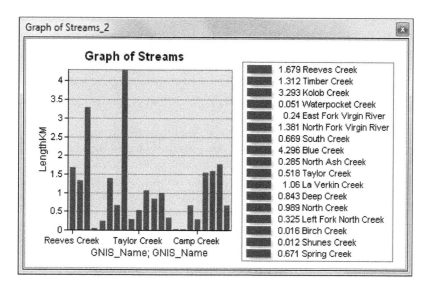

11. Right-click on the graph, and you will notice that you now have the capability of saving it as an ArcGIS graph file or exporting it to a number of different formats. This gives you lots of flexibility regarding what you want to do outside of ArcGIS.

12. However, this doesn't show you how to embed your map into an existing layout. This is pretty easy. Go to the top of your page and select the View pull-down and select Layout View.

13. Now go back to your graph, right-click, and select Add to Layout. The software immediately imports the graph to your layout, but as you notice, it's probably neither the right size nor in the right place.

14. Grab the tiny blue resizing boxes and shrink or enlarge to fit the underlying layout. Then grab the center of the object to reposition it where you like. The result should look something like the following.

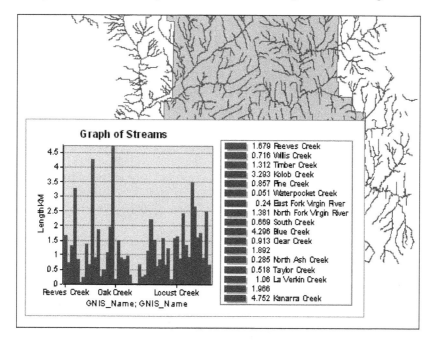

You have finished. Imagine how much flexibility this would give you in creating your presentations.

Route Logs

Those of you familiar with specialized online route-finding software might recognize the route log, which is a set of turn-by-turn directions often accompanying a map showing "from" and "to" directions. This process requires road network databases that contain a substantial amount of information about the nature of the roads, speed limits, construction, traffic flows at different times of the day, and other road information that might affect one's speed going from point A to point B. That can take a substantial amount of time to produce and possibly longer to explain. The purpose of this section is simply to show you how easy it is to generate a route log once that database is in place and an analysis has been performed.

You might ask why one needs such a device if the map is already there. The answer is pretty much the same as for any graphic added to the map—to increase its utility. Some people have a hard time reading a map, or the map might not be in a sufficiently large scale for the reader to find useful. The route log is very specific and can act alone or can augment a traveler's ability to read and understand what the map is trying to say. The review and ArcGIS exercise Activity 10-2 should help you understand why they are used and how to create these.

ACTIVITY 10-2   WHY ROUTE LOGS

This activity allows you to recall what you learned about why route logs are useful. Bring your own experiences to focus on these questions.

1. What is a route log? What does it describe?

2. If you have a map of your route from point A to point B that was produced by ArcGIS, what are some

possible reasons for wanting a route log as well as the map?

3. What are some of the things you will need in your network database for the software to conduct a shortest path analysis?

Web Maps and Story Maps

Up to now, you have focused on using ArcGIS as the primary platform for doing spatial analysis, building models, and generating output based on that work. As you have already discovered from the introductory exercises, there is an enormous amount of power included in ArcGIS, but it also has a fairly substantial learning curve that, for some, is just a bit too steep to climb. Fortunately, many of Esri's products are now going online and there is an emerging product, ArcGIS Online found at ArcGIS.com. At one time, the product was mostly for simple measurement and map display, but each month it seems to grow and get more powerful.

ArcGIS Online is a web-based GIS software environment that allows for the creation, sharing, and collaboration of geospatial data and analytics among members of an organization that possesses an account with Esri. With ArcGIS Online, you can use and create maps and scenes and access layers, analytics, and ready-to-use maps; you can publish map layers on the web so they can be accessed from any device with Internet capability; and you can convert your Microsoft Excel spreadsheet data into geospatial data, customize the ArcGIS Online website, and create location-based apps.

ArcGIS Online requires web browsers and devices compatible with the Internet. In addition, you must have direct access to ArcGIS Online from within the ArcGIS desktop, giving you the power of both environments. To access ArcGIS Online, you can use a public account as you did in ArcGIS Lesson 2-1. That account allows you to view map information and to have limited access to some of its analytical capabilities. Alternatively, a paid account is available that allows you to gain anonymous access to your organization's site where you have access to any information shared with you or with the public. These paid accounts require the purchase of credits that are determined by data and usage. This section will give you a brief overview of ArcGIS Online.

ACTIVITY 10-3 | WEB MAPS AND STORY MAPS

This activity allows you to recognize what you know about web mapping and story maps and to learn more. Take the time to answer the following questions carefully and thoughtfully.

1. Based on Esri's definition, what is a web map and what are its capabilities?

2. Go to the Esri webswite and investigate **story maps**. How does it differ from a web map? How might you use it in your work?

3. While you are at the Esri site, find out how it defines **feature service**. In your own words, how would you define it, and what are the advantages it has over regular web maps?

4. Before you do ArcGIS Lesson 10-2 go to the Esri website and gather information about what basic templates are available for story map development. List them here. What is an example of how you might use each?

ARCGIS LESSON 10-2 | MAKING WEB MAPS AND STORY MAPS

This lesson provides a quick example of how to create an Esri universal account and to get a free, 60-day ArcGIS Online account as well as how to get a paid account if you or your organization desires it. I will also show you how to create a simple web map so that you understand the power of this approach. Besides creating maps, ArcGIS Online also provides an increasing set of analytics, and you will get a chance to see what tools are available.

Besides making maps and analyzing maps online with an easy user interface, ArcGIS Online provides many opportunities to tell stores with maps by combining the maps with multimedia,

text, and narration. You will get a chance to work with the story maps in this exercise as well.

Just so you are in touch with what's happening with Esri and so you don't get dozens of accounts, it's a good idea to get a universal account with the company.

1. Go to https://accounts.esri.com/?redirect_uri=http://www.esri .com/search#get an account.

2. If you don't have an account, select Create an Esri Account. The following simple menu appears. If you already have a universal account, just skip this step.

(continued)

ARCGIS LESSON 10-2 | (CONTINUED)

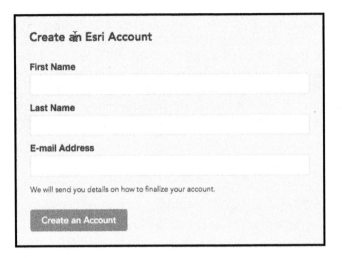

3. Whether you have a universal account or not, begin the process of creating an ArcGIS Online account by logging on to www.arcgis.com.

4. If you have an account, log in. Otherwise fill in the form to create a trial account.

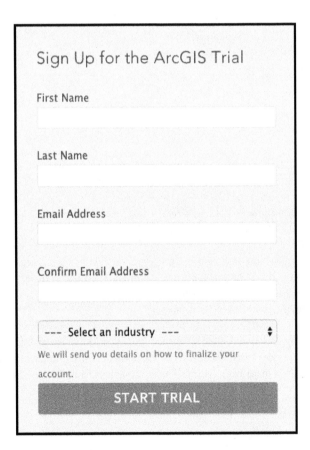

5. Once you have completed the form, you can log into your ArcGIS account. It will ask for your universal Username and Password.

6. On the bottom of the screen, one of the options is Make a Map. Click on it.

7. By default, North America appears.

8. Many of the mapping tools appear at the upper left of the screen. A description of what they do follows.

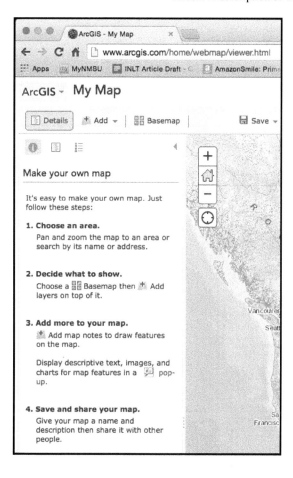

(*continued*)

ARCGIS LESSON 10-2 | (CONTINUED)

9. To zoom in, click on the plus sign as many times as you wish. To pan, move your cursor, which will appear as a hand when moving to place your map where you wish.

10. I chose to zoom in on my hometown; you might do the same. I also selected a different basemap.

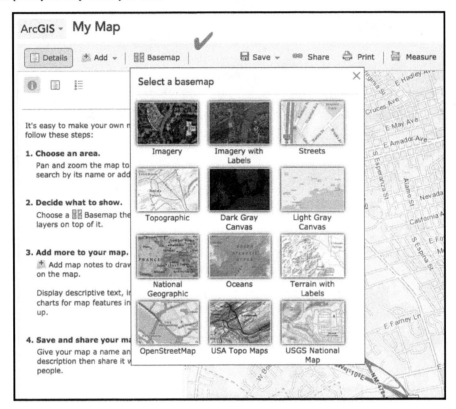

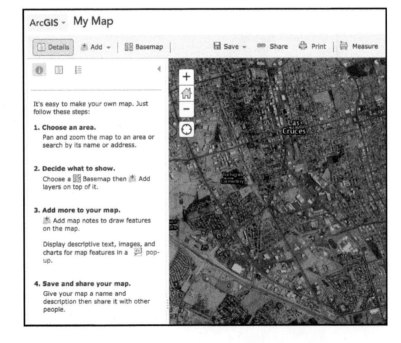

11. To add data, all you need to do is select the Add button and a number of different sources of available maps, including some you might have created, are at your fingertips.

12. I wanted to see what types of soils were located in my hometown, so I selected the USA Soil Survey layer. Notice that a description of each layer pops up when you click on it. Then the option to add the layer also appears.

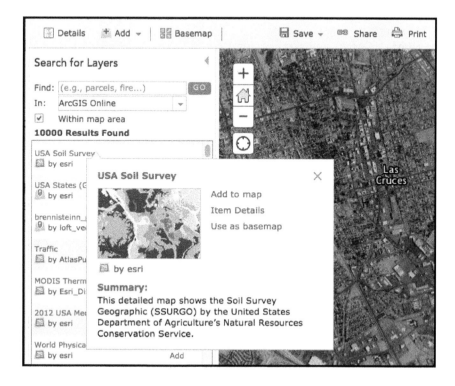

(continued)

ARCGIS LESSON 10-2 | (CONTINUED)

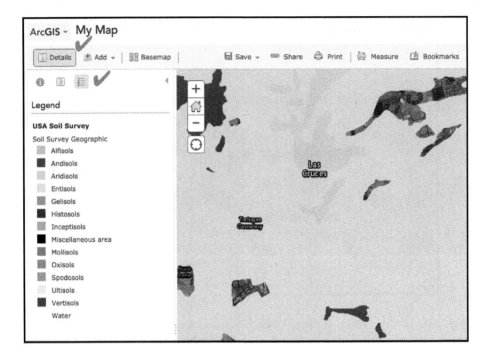

13. Now that I have the layer I want, I need to save it. The Save icon allows you not only to save your work but also to provide a title, tags for searching, a descriptive summary of your map, and the online folder (ArcGIS Online) or any other place you wish.

14. Now that you've seen how this works, go ahead and do one of your own. Turn it in with a brief description of what area you chose, why you chose it, and why you selected the layers you did.

15. You are actually very close to having produced not only your first web map but also your first story map. You know why you chose your area, what layers you wanted, and basically the story you wanted to tell. All you need now is a cool way

to present all that in one place—hence, the story map. Story maps are easy to produce. Rather than giving you a lengthy description, especially one that already exists. just do the following.

  a. Go the following URL: http://storymaps.arcgis.com/en/

  b. Look through some of the many existing story maps already there.

  c. When you finish that, click the Check out all of our apps hotlink. *Note:* You may have already found this in Activity 10-2, get a feel for what you can do with these.

16. Make a story map of the web map you created earlier or choose a new location. Tell a story. Perhaps tell your class-mates something about where you once lived. Turn this in to your instructor when finished.

ADDITIONAL READING AND RESOURCES

http://blogs.esri.com/esri/arcgis/2010/02/08/var-chart-legends/

KEY TERMS

bar graph: A graph consisting of two or more parallel bars oriented either vertically or horizontally, each representing a particular attribute value.

feature service: A web-based service that allows one to serve map features over the Internet, allowing clients to manipulate symbology, execute queries, and perform edits that can be applied to the server. One advantage is that it provides templates for clients to use for an enhanced editing experience.

histogram: A type of graph that is a subtype of the vertical bar graph that represents the frequency distribution of values.

line graph: A type of graph designed to connect data points that indicate successive changes in some measureable value or quantity.

pie chart: A circular graphic divided into wedge-shaped segments, each of which is sized to represent the proportion of the whole consistent with what that segment represents.

route log: Step-by-step text-based directions detailing a route from one point to another along a network.

story map: Web-based combination of authoritative map content, images, narration, and multimedia that work together to tell a story.

Endnote

You have by now probably read most of this text and/or completed most of the activities and lessons. This should not be the end of your learning but rather just the beginning. I urge you learners to continue to seek out any opportunity you have to continue your learning. Several opportunities for further education that you should consider as your career moves forward include additional, more advanced community college or college coursework, vendor workshops, training at professional meetings, and other nontraditional yet formal ways of learning. Beyond that I highly recommend that you join online GIS communities (e.g., some of those you found in Chapter 1) and share your experiences with those with whom you will hopefully interact for many years to come. You might have additional material provided by your instructor. Take advantage of any such material, especially if it includes projects that put all these pieces together.

Printed in the USA
K050655SCI041817   01S29053000000001893